Heat Convection in Micro Ducts

THE KLUWER INTERNATIONAL SERIES IN MICROSYSTEMS

Consulting Editor: Stephen Senturia
Massachusetts Institute of Technology

Volumes published in
MICROSYSTEMS

Optical Microscanners and Microspectrometers Using Thermal Bimorph Actuators
Gerhard Lammel, Sandra Schweizer, Philippe Renaud
ISBN 0-7923-7655-2,

Scanning Probe Lithography
Hyongsok T. Soh, Kathryn Wilder Guarini, Calvin F. Quate
ISBN 0-7923-7361-8

Microsystem Design
Stephen Senturia
ISBN: 0-7923-7246-8

Microfabrication in Tissue Engineering and Bioartificial Organs
Sangeeta Bhatia
ISBN 0-7923-8566-7

Microscale Heat Conduction in Integrated Circuits and Their Constituent Films
Y. Sungtaek Ju, Kenneth E. Goodson
ISBN 0-7923-8591-8

Micromachined Ultrasound-Based Proximity Sensors
Mark R. Hornung, Oliver Brand
ISBN 0-7923-8508-X

Bringing Scanning Probe Microscopy Up to Speed
Stephen C. Minne, Scott R. Manalis, Calvin F. Quate
ISBN 0-7923-8466-

Microcantilevers for Atomic Force Microscope Data Storage
Benjamin W. Chui
ISBN 0-7923-8358-3

Methodology for the Modeling and Simulation of Microsystems
Bartlomiej F. Romanowicz
ISBN 0-7923-8306-0

Materials & Process Integration for MEMS
Francis E.H. Tay
ISBN 1-4020-7175-2,   September 2002

Microfluidics and BioMEMS Applications
Francis E.H. Tay
1-4020-7237-6

Heat Convection in Micro Ducts

Yitshak Zohar
*The Hong Kong University
of Science and Technology*

Kluwer Academic Publishers
Boston/Dordrecht/London

Distributors for North, Central and South America:
Kluwer Academic Publishers
101 Philip Drive
Assinippi Park
Norwell, Massachusetts 02061 USA
Telephone (781) 871-6600
Fax (781) 681-9045
E-Mail: kluwer@wkap.com

Distributors for all other countries:
Kluwer Academic Publishers Group
Post Office Box 322
3300 AH Dordrecht, THE NETHERLANDS
Telephone 31 786 576 000
Fax 31 786 576 254
E-Mail: services@wkap.nl

 Electronic Services < http://www.wkap.nl>

Library of Congress Cataloging-in-Publication Data

Yitshak Zohar
Heat Convection in Micro Ducts
ISBN 978-1-4419-5320-9

Printed on acid-free paper.

Printed in the United States of America.

To my dear *NILI*

IN MEMORY
of my mother *MIRYAM*
and late brother *YAAKOV*

Contents

Foreword

Yitshak Zohar's book on *Microchannel Heat Sinks* is the second book in the Microsystems Series to focus on heat transfer in microsystems. The first, *Microscale Heat Conduction in Integrated Circuits and Their Constituent Films*, by Ju and Goodson, dealt with the problem of conduction in heterogeneous microstructures and presented new methods of characterizing the heat flow. This new volume addresses the coupling between heat conduction and fluid mechanics that is central to understanding convective heat transfer, and does so in the context of microfabricated structures.

Understanding the fundamentals of microchannel heat transfer is essential not only for the important task of heat removal in integrated circuits, but also for a growing family of microfluidic devices designed for chemical and biochemical analysis. Thus, the book includes both conventional pressure-driven fluid flow and electrokinetic transport along with applications appropriate to each type of flow. There is good balance between the fundamentals of the relevant theory and the approximations that are used to create tractable solutions to practical problems so that experimental behavior can be confronted on a sound footing.

Microchannel Heat Sinks will be a welcome addition to the microsystem engineer's bookshelf.

Stephen D. Senturia
Brookline, MA

Preface

Electronic cooling has been the technology driver for the developments of microcooling systems since the early 1980's. As electronic products become faster and incorporate greater functionality, they are also shrinking in size and weight, with continuing demand for cost reduction. Shrinking system sizes are resulting in increasing heat generation rates and surface heat fluxes leading to a significant interest in ultra-compact high heat removal devices. However, with the proliferation of microsystems to all aspects of our life, thermal control and management on a microscale is becoming a critical factor in a wide range of applications. The concept of lab-on-a-chip requires several thermal components and, ultimately, the technology is directed toward chemical and biochemical analysis tools based on the miniaturization and integration of many elements on a single chip. For commercial success, this new technology not only satisfies the need of a large demand but also promises enormous potential. This broad base application may become the prime technology driver for the development of thermal microsystems.

The traditional air-cooling method of finned heat sink with an attached dc fan is clearly approaching its limits. Therefore, the first alternative is liquid cooling with or without phase change. Most of the proposed cooling techniques utilize, of course, the latent heat of phase change for enhanced heat transfer efficiency. As the area of micro fluid mechanics and heat transfer continues to grow, it becomes increasingly important to understand the fundamental mechanisms involved with single- and two-phase micro-scale heat convection. In particular, the subject of convective boiling in microchannels is relatively young, and most of the work has been done within the last decade. The goal of this book is, therefore, to summarize the findings of this early research stage of convection heat transfer in micro ducts. However, typical to any emerging field, the understanding of fundamental principles is often neglected in favour of discovering new applications. Thus, it is inevitable that some initial concepts and preliminary results are not mature yet. Indeed, the book was written as a stepping-stone with the hope of stimulating discussion and research among colleagues to enhance our knowledge for the benefit of all.

Yitshak Zohar

Acknowledgements

I would like to express my gratitude to Professors Chih-Ming Ho of UCLA and Yu-Chong Tai of Caltech, who introduced me to the micro world a decade ago, and Professor Man Wong of HKUST, who has been working with me shoulder-to-shoulder for a decade now. The interaction with them over the years has been the source of inspiration and guidance required for a sustained research work.

I'm particularly grateful to the colleagues and students who spent many pain-stacking hours fabricating devices, testing them, and analyzing the results. The bulk of the book material is based on their work. Prof. Yuelin Wang initiated the fabrication and testing methodology of integrated thermal microsystems. Dr. Linan Jiang continued in his footsteps, and her Ph.D. thesis is the main source for much of the book content; this includes the steady and unsteady, single- and two-phase microchannel heat sink as well as the thermistor studies. Finally, Mr. Man Lee, through his M.Phil. thesis on micro heat pipes, together with Mr. Yiu Yan Wong, working on a rectangular microchannel heat sink, contributed their fare share to the evolving research program.

Several other colleagues and students who indirectly contributed to this effort are Prof. Xinxin Li, Prof. Xin Zhang, Ms. Wing Yin Lee, and Mr. Sylvanus Yuk Kwan Lee.

The unsung heroes are perhaps the members of the technical support staff of the Micro Fabrication Facility at HKUST, who quietly and diligently provide efficient and highly professional service.

Last but certainly not least is the tremendous support of the Lab Scientific Officer Mr. Wan Lap Yeung. He humbly, behind the scenes, has been welcoming into the group every newcomer, and patiently provides the proper training critically needed in the initial stage of every research project.

Special thanks are due to Dr. Linan Jiang of Stanford, who read the draft, corrected some glaring errors, and made a few significant comments and suggestions. Her efforts, no doubt, enhanced the quality of the final product.

The funding by the Research Grants Council, through the Competitive Earmarked Research Grant program, is highly appreciated. It has been instrumental in enabling us to develop the MEMS program at HKUST.

Chapter 1

Introduction

The last decade of the twentieth century has witnessed an impressive progress in micromachining technology enabling the fabrication of micron-sized devices, which become more prevalent both in commercial applications and in scientific research. These microsystems have had a major impact on many disciplines, e.g. biology, chemistry, medicine, optics, aerospace, mechanical and electrical engineering. This emerging field not only provides miniature transducers for sensing and actuation in a domain that we could not examine in the past but also allows us to venture into a research area in which the surface effects dominate most of the physical phenomena [62]. Fundamental heat-transfer problems posed by the development and processing of advanced Integrated Circuits (ICs) and MicroElectroMechanical Systems (MEMS) are becoming a major consideration in the design and application of these microsystems. The demands on heat removal and temperature control in modern devices, with highly transient thermal loads, require new techniques for providing high cooling rates and uniform temperature distributions. Thus, thorough understanding of the physical mechanisms dominating microscale heat transfer is vital for continuous evolution and progress of microdevices and microsystems.

1.1   Electronic cooling

A fundamental requirement for the commercial success of any micro-fabrication technology is an application with a very large demand. These applications are essential technology drivers providing sufficient economic pull for the adequate recovery of facility costs, which sustain continued research into new and improved devices at very low unit cost.

It is envisioned that, sooner or later, billions of people, places and systems could be all connected to each other and to useful services through the Internet. Then, scalable and cost-effective information technology capabilities will need to be provisioned, delivered, metered, managed and purchased as a service [44]. Consequently, processing and storage will be accessible via a utility. In this model, planetary-scale computing, customers will pay for services based on the amount they use, similar to other public utilities such as electricity and water. Some of the technical challenges facing the designers are thermal and mechanical issues of future integrated circuits, densely packaged racks of servers, and large scale data centres. It is estimated that a planetary-scale data centre will require 50MW of power for the computing infrastructure, and the cooling of this centre will consume an additional 25MW.

There is no denying that the prime technology driver for thermal microsystems has been electronic cooling. The picture of a modern microprocessor under a finned heat sink with an attached dc fan, like the example shown in Figure 1, is by now a household recognized image. This picture demonstrates both the problem of heat generation during the operation of microelectronic devices, and the traditional solution of heat removal using forced convection air-cooling through a network of fins. However, it is well known that this cooling technology is approaching its limits.

Figure 1.1: A picture of a modern CPU cooling system featuring a finned heat sink with an attached DC fan.

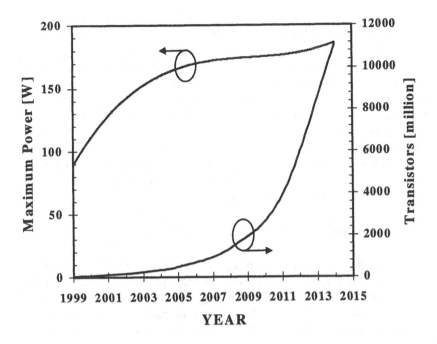

Figure 1.2: The projected number of transistors and corresponding heat generation according to the roadmap.

Indeed, the role of thermal and mechanical architecture is critical to ensure efficient power management, cooling and physical data centre design. These trends motivate a variety of thermo-mechanical research topics related to innovative thermo-mechanical designs that are required for the next generation of chips, the total energy costs over the lifetime of a product, and techniques that allow using and re-using energy more efficiently. To meet this challenge, The Defence Advanced Research Project Agency (DARPA) launched the Heat Removal By Thermo-Integrated Circuits (HERETIC) program, and a project summary is now available [35]. For the immediate future of the high-performance chip market, it is projected that by the year 2003 the microprocessor frequency will reach 3.5GHz, the device feature size will decrease to 0.1µm, the number of transistors per chip will increase to 200 million and the number of I/O connections per chip will increase to 4000. In the International Technology Roadmap for Semiconductors (2000 Update) [72], it is predicted that within a decade the number of transistors per device will reach 12×10^9 with heat generation of close to 200W as shown in Figure 1.2. Currently, all the advanced technologies have demonstrated cooling capability of less than to 100W/cm^2.

As early as two decades ago, Tuckerman and Pease [208] pioneered the use of microchannels for the cooling of planar integrated circuits. They demonstrated that by flowing water through small cooling channels etched in a silicon substrate; heat transfer rates of about $10^5 \text{W/m}^2\text{K}$ could be achieved. This rate is about two orders of magnitude higher than that in the state-of-the-art commercial technologies for cooling arrays of ICs. Since then heat transfer performance of flows in microchannels has been experimentally and theoretically analysed by a number of researchers. Most of the early studies dealt with single-phase flow of either liquid or gas through microchannels, as it proved to be complicated enough. However, it is clear that utilizing the latent heat associated with phase-change can dramatically enhance the performance of microchannel heat sinks.

1.2   Thermal issues in microsystems

Although electronic cooling is recognized as the current technology driver for thermal management solutions, similar problems exist in other microsystems. Microsystem technology generally involves coupled-field behaviour, combining two or more energy domains such as mechanical, electrical, chemical or thermal [36]. The electrical and thermal domains are probably the most prevalent in microdevices, either by design or as a by-product. Unlike thermal management in electronic packaging, concerned mainly with efficient heat removal, thermal challenges in microsystems encompass a large number of issues; ranging from harnessing thermal energy for driving microactuators to the ultimate sensor noise limits imposed by inherent thermo-mechanical coupling in microsensors.

One of the earliest and most successful applications of thermal microsystems is the microfabricated for thermal inkjet print heads [143]. Microsystem technology has been a major enabler in this field since the mid-1970s, allowing significant reduction in cost while increasing the performance of printers. The operating principles are based on a combination of thermodynamics and fluid dynamics using silicon micromachined components. In these devices, a vapour bubble is thermally generated and expanded within a silicon chamber using an integrated microheater, forcing ink to be expelled from an exit nozzle. A similar technique was applied to micro-machined fuel injectors for automotive applications, and a novel design using multiple thermal elements for rapid droplet generation was also demonstrated [204].

Bulk silicon has been extensively applied to thermal isolation of passive and active on-chip components. Isolation to remove heat sources or sinks in the proximity of a thermal element is typically achieved through etching of either the bulk substrate or a thin film sacrificial layer underneath the thermal element. Thermal isolation structures can be found in high performance thermal sensors. For example, thermoelectric infrared detector arrays have been fabricated using a commercial 1μm-linewidth CMOS process, with the polysilicon/aluminium interface acting as a thermopile detector and backside bulk micromachining providing thermal isolation [135]. Beyond passive thermal isolation, active thermal control has also been used for applications such as frequency tuning for microresonators, which often exhibit high temperature sensitivity that can limit their performance [174]. Active thermal control has been used to perform direct mechanical actuation through thermal expansion. With relatively low voltage requirements and high force output, electrothermal actuation has important advantages over traditional electrostatic actuation. Both in-plane and out-of-plane actuators have been developed that rely on resistive heating and thermal expansion of various materials [142,168].

Thermal expansion for controlled mechanical work has been applied to realize micropumps and microvalves for gas and liquid manipulation. Rather than using bubble expansion to expel fluid from the micromachined chip, as in the case of inkjet print head, a bubble pump for fluid mixing in a stationary micromachined chamber has been demonstrated [41]. Due to high surface tension forces, thermally generated microscale bubbles tend to be highly stable and useful for generating mechanical work. An extension of fluid displacement via thermally generated bubbles has also been used to generate bulk pumping in microchannel networks [233]. A valveless micro-pump was demonstrated by combining a series of addressable microheaters to create a travelling sequence of bubbles capable of precise fluid pumping with a 0.5nL/min flow rate [103]. A larger scale thermal bubble micropump capable of delivering a 6.5mL/min flow rate [203], and thermal bubble micropumps for drug dispensing from a micro-needle [234] have also been fabricated. Other micropumps utilize alternate thermal actuation methods such as bimetallic and thermopneumatic elements [241], or TiNi shape memory alloy actuators with integrated thermal control [121].

Another widely used application of thermal microsystems is the sensing of bulk fluid flow in microdomains. A variety of micromachined thermal sensors for measuring liquid flow rates and wall shear stress have been demonstrated, such as microscale hotwire anemometers fabricated from polysilicon elements [67,79], and anemometers based on p/n polysilicon thermopiles with integrated heating elements [129]. The heaters in these devices function as classical hotwire anemometers, with a feedback loop to maintain the temperature of the element as heat is removed by the flow. The sensors have demonstrated a flow rate resolution of 0.4nL/min. [225]. Such resistive Joule heating elements for anemometry can be readily formed using resistive metal or selectively doped bulk silicon.

An important application for thermally actuated pumping, valving, and flow sensing is lab-on-a-chip microsystems. The development of lab-on-a-chip technology is ultimately directed toward chemical and biochemical analysis tools based on the miniaturization and integration of liquid handling, sample processing, and sample analysis in a single chip. In addition to flow control applications, other issues in lab-on-a-chip concept are directly related to thermal phenomena. For example, microdevices for polymearase chain reaction (PCR) amplification of DNA have been developed by a number of groups. The PCR process involves sequential heating and cooling, with 30s at 95°C for denaturation of double-stranded DNA molecules, and 1-2min each for annealing and extension at 72°C to form a new set of double-stranded molecules. With 25-40 cycles per reaction, a traditional cycler requires about 90min for amplification. In contrast, micro heaters integrated in a micro reaction chamber for PCR amplification have been reported with a cycle time of 60s/cycle and heating rate of 15°C/s, for a total amplification time of approximately 22min [137]. More recently, an integrated device that accomplishes cell lysis, PCR, and capillary electrophoresis on monolithic microchips has been demonstrated using bacterial samples [216]. Rapid PCR thermal cycling rates as fast as 17 s/cycle have been demonstrated in silicon microchambers [10]. The short cycle times are made possible by the relatively small thermal mass of the microstructures. This technology is based on temporal control over the temperature field within a single reaction chamber. Similarly, spatial temperature control has been used for continuous flow PCR, based on a glass chip with a single microchannel that meanders the sample across three zones of constant temperature to provide 20 amplification cycles in times ranging from 1.5 to 18.5 min [101].

Genetic tests have an enormous scope of applications in biotechnology and medicine, ranging from agriculture and farming to genetic diagnostics on human subjects. Currently, about 400 diseases are diagnosable by molecular analysis of nucleic acids, and the number is continuously increasing. Many of these assays have been developed as part of a major thrust aimed at making medicine a more quantitative science. Furthermore, many more assays will follow as more genetic information is discovered by major research efforts such as the Human Genome project. Humans have approximately 100,000 genes that could be potentially tested for defects or the propensity for diseases. Essentially with the same procedure, the contents of every gene on any form of life could be examined. This new development of microsystem technology not only satisfies the requirement for a huge demand but also promises enormous potential for growth. Such a broad base application may indeed prove to be the ultimate technology driver of all time.

1.3   Case studies

As the limit for air-cooling (finned heat sink with a dc fan) is reached, the first alternative is liquid cooling with or without phase change. So far, liquid-cooled electronics has not penetrated the market, with the exception of heat pipes used in laptop computers. Intense activities are currently taking place, such as arrays of micro impinging jets with the advantage of single-phase operation [227]. The majority of the proposed cooling techniques, though, utilize the latent heat of phase change for enhanced efficiency. The electrokinetic microcooler incorporates an electro-osmotic pump in conjunction with a microchannel heat sink for a continuous, closed-loop operation [78]. A micro capillary pumped loop, comprising an evaporator, condenser, reservoir and fluid lines, is capable of carrying a much greater heat load due to its single directional flow [165]. A compact thermosyphon loop includes an array of channels and pores fabricated inside the evaporator section to enhance the boiling performance [172]. Other techniques such as cross-flow micro heat exchangers [57], spray cooling [139] or thermoelectric microcoolers [232] have also been studied in association with thermal management of microsystems. Indeed, a diversity of ideas.

However, as the area of micro fluid mechanics and micro heat transfer continues to grow, it becomes increasingly important to understand the mechanisms and fundamental differences involved with heat transfer in single- and two-phase flows in micro ducts. The subject of two-phase forced convection heat transfer in microchannels is relatively young, and most of the work has been carried out within the last decade. By far, the majority of the reported research work in this area has been empirical in nature [86]. Only recently, efforts to derive analytical models from basic principles rather than empirical correlations have been reported [146].

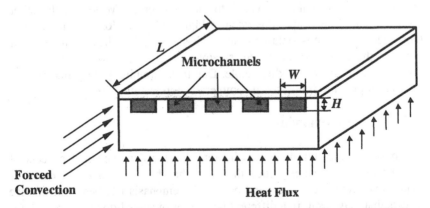

Figure 1.3: A schematic illustration of a microchannel heat sink showing the heat flux from the source and the microchannel forced convection for heat removal.

Heat input   Heat output

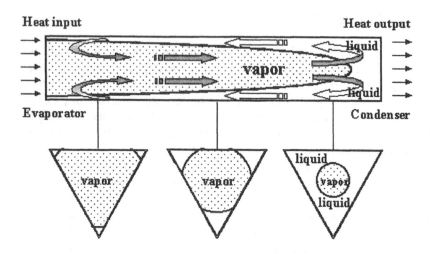

Figure 1.4: A schematic illustration of a micro heat pipe showing the evaporator where heat enters and the condenser where heat is rejected from the system together with the flow pattern and the corresponding cross-sections.

Two examples of thermal microsystems have been selected as case studies for a detailed discussion throughout the book: (i) microchannel heat sink, and (ii) micro heat pipe. The two microsystems are fundamentally different, yet both feature phase change and consequently two-phase flows. The microchannel heat sink, schematically illustrated in Figure 1.3, is an open system. Liquid is forced through the microchannel under pressure gradient, which is imposed by an external power source such as a pump. Even if all the control parameters are constant in time, the developing flow could be highly irregular. Although very promising, due to its complexity, this approach is yet to make in roads into the commercial world. On the other hand, the micro heat pipe illustrated in Figure 1.4 is a closed system. Once fabricated, it is self-contained and self-started. The heat source itself provides the driving force. Under steady-state operation, within the system limitations, the flow is highly stable. Due to its simplicity, the concept of a micro heat pipe has found some commercial applications already.

1.4   Book overview

Following the Introduction, Chapter 2 reviews fundamental aspects of convective heat transfer. Since this topic has been discussed in detail in many text-, reference- and hand-books, the emphasis has been put on size effects that may result in a different behaviour of scaled-down microsystems compared to their macosystems counterparts.

Chapter 3 analyses the physics of scaling relevant to heat convection in microchannels, and introduces a host of non-dimensional parameters that are useful for the analysis. A dimensional analysis for a thermal microsystem is carried out, and the control parameters unique to microscale convective heat transfer are identified.

Chapter 4 details the major steps in the fabrication of the two case studies, i.e. microchannel heat sinks and micro heat pipes. The fabrication techniques of thermal microsystems are fundamentally the same as those used to fabricate MEMS devices, which are extensions of the IC fabrication technology. In-depth discussion of micro fabrication technologies is the subject of many books and, therefore, will not be duplicated. However, the application of such techniques to the fabrication of integrated thermal microsystems often requires an innovative approach.

Chapter 5 is dedicated to thermometry techniques that have been applied for the characterization of thermal microsystems. It is becoming evident that in scaling down a thermal microsystem, the main challenge often lies in the characterization of its performance and not so much in its fabrication. Indeed, in many cases, reliable experimental data with high spatial and temporal resolution can be obtained only by integrating microsensors for measurements. In particular, the design, fabrication and analysis of integrated microsensors for temperature and void-fraction measurements are presented.

Chapter 6 presents experimental and theoretical studies of steady-state, single-phase heat convection in micro ducts. After presenting the basic physical concepts related to single-phase fluid flows, the discussion is separated into gas and liquid flows, which exhibit fundamentally very different size effects.

Chapter 7 deals with steady state, two-phase convective boiling in micro ducts. The onset of bubble formation and subsequent bubble dynamics, which are the early symptoms of phase change, have been the subject of intense debate. The geometric size effect on these processes is not as yet clear, and more work is needed to clarify these issues.

Chapter 8 concentrates on the unsteady operation of thermal micro-systems. In Chapters 6 and 7, the microsystem control parameters, such as the driving pressure difference or input heat flux, are constant in time. However, the resulting flow and thermal fields could very well be unsteady, especially in convective boiling as demonstrated in Chapter 7. In contrast, in Chapter 8, the transient performance of thermal microsystems due to time-dependent heating is presented; in particular, the thermal response to step and periodic functions of the input power is discussed.

Chapter 9 presents the concept, fundamentals as well as steady-state and transient operation of micro heat pipes. A micro heat pipe is a closed system, drastically different from a microchannel heat sink that is an open system; hence, much easier to study theoretically and experimentally. Although the maximum heat transport capacity of micro heat pipes is smaller compared to microchannel heat sinks, micro heat pipes have a tremendous advantage being self-contained and self-started. Indeed, this class of devices has already found its place in commercial applications. The book ends with a list of references.

Chapter 2

Fundamentals of Convective Heat Transfer in Micro Ducts

In the science of thermodynamics, which deals with energy in its various forms and with its transformation from one form to another, two particularly important transient forms are defined: work and heat. These energies are termed transient since, by definition, they exist only when there is an exchange of energy between two systems or between a system and its surrounding. When such an exchange takes place without the transfer of mass to/from the system and not by means of a temperature difference, the energy is said to be transferred through the performance of work. If the exchange of energy between the systems is the result of temperature difference, the exchange is said to be accomplished via the transfer of heat. The existence of a temperature difference is the distinguishing feature of the energy exchange form known as heat transfer. Microchannel heat sinks or micro heat pipes are a class of devices that can be applied for the transfer of thermal energy from very small areas.

2.1   Modes of heat transfer

The mechanism by which heat is transferred in an energy conversion system is complex. However, three basic and distinct modes of heat transfer have been classified: conduction, convection and radiation [177]. Convection is the heat transfer mechanism, which occurs in a fluid by the mixing of one portion of the fluid with another portion due to gross movements of the mass of fluid. Although the actual process of energy transfer from one fluid particle or molecule to another is still heat conduction, but the energy may be transported from one point in space to another by the displacement of the fluid itself. An analysis of convective heat transfer is, therefore, more involved than that of heat transfer by conduction alone since the motion of the fluid must be studied simultaneously with the energy transfer process. The fluid motion may be caused by external mechanical means, e.g. pumps, in which case the process is called forced convection. If the fluid motion is caused by density differences created by the temperature differences existing in the fluid mass, the process is termed free or natural convection. The important heat transfer in liquid-vapour phase change processes, i.e. boiling and condensing, are also classified as convective mechanisms, since fluid motion is still involved with the additional complication of a latent heat exchange [190]. Hence, heat transfer in microchannel heat sinks and micro heat pipes belongs to this class of forced convection heat transfer with or without phase change.

2.2   The continuum hypothesis

In an analysis of convective heat transfer in a fluid, the motion of the fluid must be studied simultaneously with the heat transport process. In its most fundamental form, the description of the motion of a fluid involves a study of the behaviour of all discrete particles, e.g. molecules, which make up the fluid. The most fundamental approach in analysing convective heat transfer would be, therefore, to apply the laws of mechanics and thermodynamics to each individual particle, or a statistical group of particles, subsequent to some initial conditions. Such an approach, kinetic theory or statistical mechanics, would give an insight into the details of the energy transfer processes; however, it is not practical for most scientific problems and engineering applications.

In most applications, the primary interest lies not in the molecular behaviour of the fluid, but rather in the average or macroscopic effects of many molecules. It is these macroscopic effects that we ordinarily perceive and measure. In the study of convective heat transfer, therefore, the fluid is treated as infinitely divisible substance, a continuum, while the molecular structure is neglected. The continuum model is valid as long as the size and the mean-free-path of the molecules are small enough compared with other dimensions existing in the medium such that a statistical average is meaningful.

However, the continuum assumption breaks down whenever the mean-free-path of the molecules becomes on the same order of magnitude as the smallest significant dimension of a given system. In gas flows, the deviation of the state of the fluid from continuum is represented by the Knudsen number defined as: $Kn \equiv \lambda / L$. The mean-free-path, λ, is the average distance travelled by the molecules between successive collisions, and L is the characteristic length scale of the flow. The appropriate flow and heat transfer models depend on the range of the Knudsen number, and a classification of the different gas flow regimes is as follows [179]:

$$Kn < 10^{-3} \qquad \text{continuum flow}$$
$$10^{-3} < Kn < 10^{-1} \qquad \text{slip flow}$$
$$10^{-1} < Kn < 10^{+1} \qquad \text{transition flow}$$
$$10^{+1} < Kn \qquad \text{free-molecular flow}$$

In the slip flow regime, the continuum flow model is still valid for the calculation of the flow properties away from solid boundaries. However, the boundary conditions have to be modified to account for the incomplete interaction between the gas molecules and the solid boundaries. Under normal conditions, Kn is less than 0.1 for most gas flows in microchannel heat sinks with a characteristic length scale on the order of $1\mu m$. Therefore, only slip flow regime, but neither transition nor free-molecular flow regime, will be discussed. The continuum assumption is of course valid for liquid flows in micro ducts.

2.3   Thermodynamic aspects (equilibrium)

The most convenient framework within which heat transfer problems can be studied is the system, which is a quantity of matter not necessarily constant, contained within a boundary. The boundary can be physical, partly physical and partly imaginary, or wholly imaginary. The physical laws to be discussed are always stated in terms of a system. A control volume is any specific region in space across the boundaries of which mass, momentum and energy may flow, and within which mass, momentum and energy storage may take place, and on which external forces may act. The complete definition of a system or a control volume must include, implicitly at least, the definition of a coordinate system, since the system may be moving or stationary. The characteristic of a system of interest is its state, which is a condition of the system described by its properties. A property of a system can be defined as any quantity that depends on the state of the system and is independent of the path, i.e. previous history, by which the system arrived at the given state. If all the properties of a system remain unchanged, the system is said to be in an equilibrium state.

A change in one or more properties of a system necessarily means a change in the state of the system has occurred. The path of the succession of states through which the system passes is called the process. When a system in a given initial state goes through a number of different changes of state or processes and finally returns to its initial state, the system has undergone a cycle. The properties describe the state of a system only when it is in equilibrium. If no heat transfer takes place between any two systems when they are placed in contact with each other, they are said to be in thermal equilibrium. Any two systems are said to have the same temperature if they are in thermal equilibrium with each other. Two systems that are not in thermal equilibrium have different temperature, and heat transfer may take place from one system to the other. Therefore, temperature is a property, which measures the thermal level of a system.

When a substance exists as part liquid and part vapour at a saturation state, its quality is defined as the ratio of the mass of vapour to the total mass, while its void fraction is defined as the ratio of the volume of vapour to the total volume. Hence, the quality and the void fraction are properties ranging between 0 and 1, and both have meaning only when the substance is in a saturated state, i.e. at saturated pressure and temperature. The amount of energy that must be transferred in the form of heat to a substance held at constant pressure in order that a phase change occurs is called the latent heat. It is the change in enthalpy of the substance at the saturated conditions of the two phases. The heat of vaporization, i.e. boiling, is the heat required to completely vaporize a unit mass of saturated liquid.

2.4   General laws

The science of heat transfer is based upon both theory and experiment. As in other engineering fields, the theoretical part is based on physical laws. Physical laws are not derived or proved from basic principles, but are simply statements based on observations of many experiments. If experimental results are found to violate a certain law, either the law must be revised or additional conditions must be placed on the applicability of the law. A physical law is called a general law if its application is independent of the medium under consideration. Otherwise, it is called a particular law. The four general laws, upon which all the analyses concerning heat transfer depend, are:

  Conservation of mass
  Conservation of momentum
  The first law of thermodynamics (conservation of energy)
  The second law of thermodynamics

The general laws, when referred to a system, can be written in either an integral or a differential form. The integral form is useful for the analysis of the gross behaviour of the flow field. However, detailed point-by-point knowledge of the flow field can be obtained only from the equations in differential form. Microchannel heat sinks and micro heat pipes typically incorporate arrays of elongated microchannels varying in cross-sectional shape. Therefore, it is most convenient to express the general laws either in rectangular or cylindrical coordinate system.

2.4.1   Conservation of mass

The law of conservation of mass simply states that, in the absence of any mass-energy conversion, the mass of the system remains constant. Thus, in the absence of source or sink, the rate of change of mass in the control volume (C.V.) is equal to the mass flux through the control surface (C.S.):

$$\frac{\partial}{\partial t} \int\limits_{CV} \rho d\upsilon + \int\limits_{CS} \rho (\mathbf{U} \cdot \mathbf{n}) dA = 0 \qquad (2.1)$$

The surface integral can be transformed into a volume integral using Gauss theorem. Since the law is valid for any arbitrary control volume, the integrand must be zero everywhere yielding:

$$\frac{D\rho}{Dt} + \rho (\nabla \cdot \mathbf{U}) = 0 \qquad (2.2)$$

This is the continuity equation, where ρ and $\mathbf{U}$ are the density and velocity vector, respectively, and D/Dt denotes the substantial derivative.

2.4.2   Conservation of momentum

Newton's second law of motion states that the sum of the external forces, $\Sigma\mathbf{F}$, acting on a system in an inertial coordinate system is equal to the time rate of change of the total linear momentum of the system:

$$\frac{\partial}{\partial t} \int\limits_{CV} \rho \mathbf{U} d\upsilon + \int\limits_{CS} \rho \mathbf{U} (\mathbf{U} \cdot \mathbf{n}) dA = \sum \mathbf{F} \qquad (2.3)$$

The external forces are classified into two categories: (i) body forces such as gravity, and (ii) surface forces such as friction. Invoking Gauss theorem in conjuction with the continuity equation, the momentum vector equation can be written as:

$$\rho \frac{D\mathbf{U}}{Dt} = \rho \mathbf{f} - \nabla P + \nabla \cdot \tau_{i,j} \qquad (2.4)$$

where $\mathbf{f}$ is the body force, P is the hydrostatic pressure, and $\tau_{i,j}$ is the viscous-stress tensor.

2.4.3   First law of thermodynamics (conservation of energy)

The first law of thermodynamics, which is a particular statement of conservation of energy, states that the rate of change in the total energy of a system, E, undergoing a process is equal to the difference between the rate of heat transfer to the system, Q, and the rate of work done by the system, W, such that:

$$\frac{\partial}{\partial t}\int_{CV}\rho e dv + \int_{CS}\rho e(\mathbf{U}\cdot\mathbf{n})dA = \frac{dE}{dt} = \dot{Q} - \dot{W} \qquad (2.5)$$

The specific energy e includes kinetic, potential and internal energy. In the absence of heat source or sink and neglecting the work done by compression, the energy equation can be written in terms of the temperature T as follows:

$$\rho c_p \frac{DT}{Dt} = k\nabla^2 T + \Phi \qquad (2.6)$$

where Φ is viscous dissipation; c_p and k are the fluid specific heat and thermal conductivity, respectively.

2.4.4   Second law of thermodynamics (entropy)

The second law of thermodynamics leads to the introduction of entropy, S, as a property of the system. It states that the rate of change in the specific entropy, s, is either equal or larger than the rate of heat transfer to the system divided by the system temperature during the heat transfer process.

$$\frac{\partial}{\partial t}\int_{CV}\rho s dv + \int_{CS}\rho s(\mathbf{U}\cdot\mathbf{n})dA \geq \int_{CS}\frac{1}{T}\frac{\delta Q}{dt} \qquad (2.7)$$

Even in cases where entropy calculations are not of interest, the second law of thermodynamic is still important as it is equivalent to stating that heat cannot pass spontaneously from a lower- to a higher-temperature reservoir.

2.5   Particular laws

The general laws alone are not sufficient to solve heat transfer problems, and certain particular laws have to be considered in the analysis. Three such laws specifically regarding heat transfer processes: conduction, convection and radiation are of interest.

2.5.1   Heat conduction

Fourier's law of heat conduction, based on the continuum concept, states that the heat flux due to conduction in a given direction, i.e. the heat transfer rate per unit area, within a medium (solid, liquid or gas) is proportional to temperature gradient in the same direction, namely:

$$\mathbf{q}'' = -k\nabla T \qquad (2.8)$$

where $\mathbf{q}''$ is the heat flux vector, and k is the proportionality constant known as the thermal conductivity of the medium under consideration. The minus sign defines the heat flow to be positive when it is in the direction of a negative temperature gradient.

2.5.2   Heat convection

Newton's law of cooling states that the heat flux from a solid surface to the ambient fluid by convection, q'', is proportional to the temperature difference between the solid surface temperature, T_w, and the fluid free-stream temperature, T_∞, as follows:

$$q'' = h\left(T_w - T_\infty\right) \qquad (2.9)$$

where h is the heat transfer coefficient. If the fluid motion involved in the process is induced by external means, such as a pump, then the process is referred to as forced convection. If the fluid motion is caused by any body force within the system, such as density gradient, the process is then called natural (or free) convection. Certain convective heat transfer processes may involve latent heat due to phase change such as boiling and condensation.

2.5.3   Heat radiation

All substances emit energy in the form of electromagnetic waves as a result of their temperature, i.e. thermal radiation, and are also capable of absorbing such energy. An ideal body that absorbs all the impinging radiation energy without reflection and transmission is called blackbody. The Stefan-Boltzmann law of radiation states that the total emission of radiation from a blackbody is related to its absolute temperature as follows:

$$q'' = \sigma_b T^4 \qquad (2.10)$$

where σ_b is the Stefan-Boltzmann constant.

2.5.4   Equation of state

A system in thermodynamic equilibrium is incapable of spontaneous change as it is in complete balance with the surroundings. The state of a pure substance at equilibrium is defined by two independent properties. All other properties are then uniquely determined in terms of these two properties. They may be found either by experiment or by the used of suitable equations of state. For example, it has been established that the behaviour of gases at low density is closely given by the ideal-gas equation of state:

$$P = \rho RT \qquad\qquad\qquad\qquad (2.11)$$

where R is the specific gas constant. At very low-density, all gases and vapours approach ideal-gas behaviour. However, the behaviour may deviate substantially from that at higher densities.

2.6   Size effects

Length scale is a fundamental quantity that dictates the type of forces or mechanisms governing physical phenomena. Miniaturization of systems, therefore, involves not new physical laws or forces but rather different physical behaviour due to the difference in relative contribution of various forces or mechanisms. The large surface-to-volume ratio in micro devices accentuates the role of surface effects. Furthermore, the intrinsic length scale in gases, the mean free path, is much larger than the average distance between liquid molecules. Consequently, in microscale convection heat transfer, the size effects observed in single-phase gas or liquid flow and two-phase vapour/liquid flow are radically different from each other.

2.6.1   Non-continuum mechanics

The flow of a gas in the continuum regime can be thought of as the combination of a random molecular motion superposed on an ordered directional flow. Near a solid boundary, there are two possible ways in which a molecule may reflect from a stationary surface as illustrated in Figure 2.1. If the solid surface is perfectly smooth, the reflection is specular; the angle of incidence is equal to the angle of reflection with no change in the velocity component tangential to the surface. In a real case, however, the surface is not smooth but rather rough, and the reflection is diffuse; the molecules may move away from the wall at any angle. For the diffuse reflection, on the average, the molecules essentially lose their tangential velocity component [4].

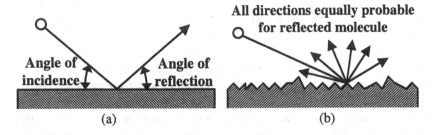

Figure 2.1: Schematic illustration of (a) specular, and (b) diffuse reflection.

If the mean free path becomes very large, intermolecular collisions become infrequent. Molecules arriving at a surface from a free stream may not have had a chance to be slowed down by collisions with other molecules; rather they arrive at the surface with free stream velocity, little affected by the presence of the solid surface. On the other hand, the molecules reflected diffusely from the wall have no net tangential velocity. Those reflected specularly have the same tangential velocity as the incident molecules. If σ_T denotes the fraction of diffusely reflected molecules, the average velocity at the surface is modelled through Maxwell's velocity-slip condition [14]:

$$U_s - U_w = \frac{2-\sigma_U}{\sigma_U}\lambda\frac{\partial U}{\partial n}\bigg|_w \qquad (2.12\text{-a})$$

Similarly, the gas temperature at the solid surface is given by Smoluchowski's temperature-jump condition:

$$T_j - T_w = \frac{2-\sigma_T}{\sigma_T}\frac{2c_p}{c_p+c_v}\frac{k}{\mu c_p}\lambda\frac{\partial T}{\partial n}\bigg|_w \qquad (2.12\text{-b})$$

U_w and T_w are the solid-surface velocity and temperature, while U_s and T_j are the gas-flow velocity and temperature at the boundary, respectively; n is the direction normal to the surface, while c_p and c_v are the specific heats. σ_U and σ_T are the tangential momentum and energy accommodation coefficients, which model the momentum and energy exchange of the gas molecules impinging on the solid boundary. They can vary from zero (specular accommodation) to one (complete, or diffuse accommodation). When gas molecules impinge on a surface, in general, the molecules exchange three properties with the surface: energy, normal momentum and tangential momentum. The extent to which impinging molecules reach equilibrium with the surface is represented by the accommodation coefficients.

The accommodation coefficients are empirically determined, and depend on both the gas molecules and the solid surface. Among the important parameters determining the reflection characteristics of a solid-gas interface is the surface roughness. If the surface is smooth on a molecular scale of the impinging gas, the reflection will be perfectly specular with no streamwise momentum transfer to the solid surface; this implies that the incident is equal to the reflected streamwise velocity and $\sigma_U = 0$. If the surface is rough on a molecular scale, all the streamwise momentum will be transferred to the surface. This diffuse reflection is characterized by complete transfer of streamwise momentum with $\sigma_U = 1$. The accommodation coefficients depend not only on the geometry of the surface but also on the molecular species adsorbed on the surface. For example, it was reported that by partially removing the adsorbed gas layer in polished steel, with surface roughness of $0.1 \mu m$, the tangential accommodation coefficient for helium could be reduced from near unity to about 0.8 [115]. However, by partially removing the adsorbed gasses, no similar reduction could be obtained for steel that was roughened. Although the exact nature of the accommodation coefficients is still an active research problem, almost all evidence indicates that for most gas-solid interactions the coefficients are approximately 1.0.

2.6.2   Electrokinetics

Most solid surfaces are likely to carry electrostatic charge, i.e. an electric surface potential, due to broken bonds and surface charge traps. When a liquid containing small amount of ions is forced through a microchannel under hydrostatic pressure, the solid-surface charge will attract the counter-ions in the liquid to establish an electric field. The arrangement of the electrostatic charges on the solid surface and the balancing charges in the liquid is called the Electric Double layer, EDL, as illustrated in Figure 2.2. Counter-ions are strongly attracted to the surface forming a compact layer, called the Stern layer, of immobile counter ions at the solid/liquid interface due to the surface electric potential. Outside this layer, the ions are affected less by the electric field and are mobile. The distribution of the counter-ions away from the interface decays exponentially within the diffuse double layer, called the Gouy-Chapman layer, with a characteristic length inversely proportional to the square root of the ion concentration in the liquid. The EDL thickness ranges from a few up to several hundreds of nanometers, depending on the electric potential of the solid surface, the bulk ionic concentration and other properties of the liquid. Consequently, EDL effects can be neglected in macrochannel flow. In microchannels, however, the EDL thickness is often comparable to the characteristic size of the channel, and its effect on the fluid flow and heat transfer may not be negligible [231].

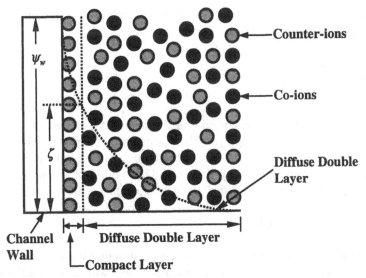

Figure 2.2: A schematic representation of an electric double layer (EDL) at the channel wall.

Consider a liquid between two parallel plates, separated by a distance H, containing positive and negative ions in contact with a planar positively charged surface. The surface bears a uniform electrostatic potential ψ_0, which decreases with the distance from the surface. The electrostatic potential, ψ, at any point near the surface is governed by the Debye-Huckle linear approximation [122]:

$$\frac{d^2\psi}{dy^2} = \frac{\psi}{\lambda_D^2} \qquad (2.13)$$

The characteristic thickness of the EDL is the Debye length, defined as

$$\lambda_D = \left(\frac{\varepsilon\varepsilon_0 k_b T}{2n_0 z^2 e^2}\right)^{1/2} \qquad (2.14)$$

where ε is the dielectric constant of the medium, ε_0 is the permittivity of vacuum, and k_b is the Boltzmann constant. z is the valence of negative and positive ions, e is the electron charge, and n_0 is the ionic concentration. If the electrical potential of the channel surface is small and the separation distance between the two plates is larger than the thickness of the EDL ($H/2\lambda_D>1$) so that the EDLs near the two plates do not overlap, the appropriate boundary conditions are: $\psi=0$ at the mid-point ($y=0$) and $\psi=\psi_w$ on both walls ($y=\pm H/2$). The solution is then:

$$\psi = \varsigma \frac{\left|\sinh\left(y/\lambda_D\right)\right|}{\sinh\left(H/2\lambda_D\right)} \tag{2.15}$$

where ς is the zeta electric potential at the boundary between the diffuse double layer and the compact layer. The potential in the diffuse double layer results in an electric body force acting on fluid particles along with the pressure and viscous forces. This force could be significant in microchannel liquid flow, and should be included in the momentum equation.

2.6.3   Polar mechanics

In addition to the usual concepts of classical nonpolar fluid mechanics, there are two main physical concepts associated with polar mechanics: couple stresses and internal spin [191]. In nonpolar mechanics, the mechanical action of one part of a body on another is assumed to be equivalent to a force distribution only. However, in polar mechanics, the mechanical action is assumed to be equivalent to both a force and a moment distribution as illustrated in Figure 2.3. Thus, the state of stress at a point in nonpolar mechanics is defined by a symmetric second order tensor, which has six independent components. On the other hand, in polar mechanics, the state of stress is determined by a stress tensor and a couple stress tensor.

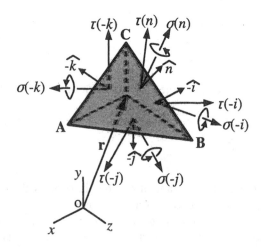

Figure 2.3: Stresses acting on an infinitesimal tetrahedron.

In classical fluid mechanics, all the kinematic parameters are assumed to be determined once the velocity field is specified. Thus, if the velocity field is identically zero, both the linear and angular momentum of all material elements must also be identically zero. However, when a particle has no translation velocity, so that its linear momentum is zero, it may still be spinning about an axis, giving rise to an angular momentum. Thus, angular momentum can still exist even in the absence of linear momentum. In polar mechanics, specifying the velocity field is not sufficient, and additional kinematic measures must be introduced to describe this internal spin.

The most important effect of polar mechanics is the introduction of a size-dependent effect that is not predicted by the classical nonpolar theories. The class of theories in which both the effects of couple stresses and micro-rotation are simultaneously taken into account in a systematic manner are termed micro fluids. Micropolar fluids are the simplest subclass of micro fluids, obtained by restricting the gyration tensor to be skew-symmetric. In micropolar fluids, rigid particles contained in a small volume can rotate about the center of the volume element described by the micro-rotation vector, but neither micro-deformation nor micro-stretch is admissible. This local rotation of the particles is in addition to the usual rigid body motion of the entire volume element. In micropolar fluid theory, the laws of classical continuum mechanics are augmented with additional equations that account for conservation of microinertia moments. Physically, micropolar fluids represent fluids consisting of rigid, randomly oriented particles suspended in a viscous medium, where the deformation of the particles is ignored. The modified momentum, angular momentum and energy equations are:

$$\rho \frac{DU}{Dt} = \nabla \cdot \tau + \rho f \qquad (2.15)$$

$$\rho I \frac{D\Omega}{Dt} = \nabla \cdot \sigma + \rho g + \tau_x \qquad (2.16)$$

$$\rho c_p \frac{DT}{Dt} = k\nabla^2 T + \tau : (\nabla U) + \sigma : (\nabla \Omega) - \tau_x \cdot \Omega \qquad (2.17)$$

where Ω is the micro-rotation vector and I is the associated micro-inertia coefficient; f and g are the body and couple force vector per unit mass; τ and σ are the stress and couple-stress tensors. $\tau : (\nabla U)$ is the dyadic notation for $\tau_{ji} U_{i,j}$, the scalar product of τ and ∇U. If $\sigma = 0$ and $g = \Omega = 0$, then the stress tensor τ reduces to the classical symmetric stress tensor, and the governing equations reduce to the classical model [117].

In micropolar fluids, the motion is affected by (i) viscous action, measure by the shear viscosity coefficient μ, (ii) microrotation, measured by the vortex viscosity coefficient κ, and (iii) the effect of couple stresses, measured by spin-gradient viscosity coefficient γ. These are material properties that give rise to a length scale l:

$$l^2 = \frac{\gamma}{\mu}\left(\frac{\mu+\kappa}{\kappa}\right) \tag{2.18}$$

Although the micropolar fluid theory has been applied to numerous problems, the primary drawback of this anlysis is the unknown viscosity coefficients γ and κ and, consequently, the polarity length scale l. Hence, further work must be done in quantifying the microstructural parameters before the theory van be fully utilized.

Chapter 3

Scaling, Similarity and Dimensionless Parameters in Convective Heat Transfer

The continuity, momentum and energy differential equations derived based on the fundamental conservation laws provide a comprehensive description of convective heat transfer. These equations, however, are so complicated that they present insurmountable mathematical difficulties. Very few exact solutions to these equations have been found, which represent very simple flow systems. Hence, the development of convective heat transfer has depended heavily on experimental research, and solutions of real problems usually involve a combination of analytical and experimental information.

3.1   Physics of scaling

Forces driving physical phenomena can be classified into two general categories: body and surface forces. Body forces depend on the third power of the length scale ($\propto L^3$), e.g. gravity, while surface forces depend on the first or the second power of the characteristic scale ($\propto L^1$ or L^2), e.g. friction. Due to the difference in slopes, body forces must intersect surface forces at some point as illustrated in Figure 3.1. Experimental observations in biological studies indicated that 1mm is the approximate crossover length scale. Experience accumulated thus far in the fabrication and operation of microsystems also suggests that surface forces dominate in length scales smaller than a millimetre [62].

A simple example illustrating this crossover length scale is the balance of forces over a particle suspended from a solid boundary due to surface tension as sketched in Figure 3.2. The gravitational body force pulling down the particle, having density ρ, is given by:

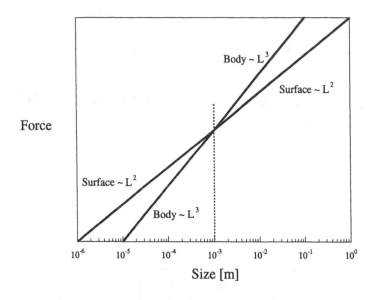

Figure 3.1: An illustration of the body and surface force dependence on size with a crossover length scale of 1mm due to the different slopes.

$$F_b = \frac{4}{3}\pi r^3 \rho g \qquad\qquad (3.1)$$

The surface tension force holding the particle in place, assuming the interface diameter is on the order of the particle size and the direction of the surface-tension force is opposite to the gravitational force, is given by

$$F_s = 2\pi r \sigma \qquad\qquad (3.2)$$

σ is the surface tension coefficient. The two forces balance each other at equilibrium and, thus, the critical size is given by:

$$r = \sqrt{3\sigma / 2\rho g} \qquad\qquad (3.3)$$

Applying Equation (3.3) to a water-solid particle system, with the solid having a density four times that of water, results in a crossover length scale of 1.5mm.

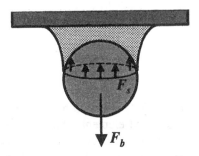

Figure 3.2: A schematic of a suspended solid particle where the gravitational force on the particle is balanced by the surface tension exerted by the liquid.

3.1.1   Length scales

Length scale is a fundamental quantity that dictates the type of forces governing certain physical phenomena. Most physical quantities scale differently with the length dimension. For isometric scaling:

$$
\begin{array}{lll}
\text{Length} & \propto L^1 \\
\text{Area} & A \propto L^2 \\
\text{Volume} & V \propto L^3
\end{array}
$$

while other forces and mechanisms scale as follows:

Time: l^0	Friction: l^2
van der Waals: $l^{1/4}$	Thermal losses: l^2
Diffusion: $l^{1/2}$	Piezoelectricity: l^2
Diffusion: $l^{1/2}$	Piezoelectricity: l^2
Distance: l^1	Mass: l^3
Velocity: l^1	Gravity: l^3
Surface tension: l^1	Magnetics: l^3
Electrostatic force: l^2	Torque: l^3
Muscle force: l^2	Power: l^3

In geometrical similarity, all dimensions are scaled by the same factor. However, for proper scaling of physical phenomena, dynamic similarity is required on top of geometrical similarity, where certain non-dimensional parameters are the same. The important dimensionless parameters can be determined by applying the well-known Buckingham Pi theorem.

3.1.2   Scaling laws

Trimmer introduced an elegant method to express different scaling laws by using a vertical bracket notation [200]. The different forces listed above can be written as one-column matrix:

$$F = \begin{bmatrix} l^1 \\ l^2 \\ l^3 \\ l^4 \end{bmatrix}$$

(3.4)

The top element in Equation (3.4) refers to a case where the force scales linearly with the size (l^1), the next element refers to a case where the force scales with the square of the size (l^2), etc. The convenience in using this represenatation can be demonstrated with the following example: as a system becomes smaller, the scaling of the force also determines the acceleration, a, the response time, τ, the amount of power generated, P, and the power density, P/V. Since the mass of the system m scales as (l^3), for a generalized case with a force scaling as (l^F), one obtains:

$$\boxed{a = F/m = \left[l^F \right] \cdot \left[l^{-3} \right]}$$

(3.5-a)

$$\boxed{\tau = \sqrt{2x/a} = \left\{ \left[l^1 \right] \cdot \left[l^3 \right] \cdot \left[l^{-F} \right] \right\}^{1/2}}$$

(3.5-b)

$$\boxed{P = Fx/\tau = \left[l^F \right] \left[l^1 \right] \left[l^{4-F} \right]^{1/2}}$$

(3.5-c)

$$\boxed{P/V = Fx/\tau V = \left[l^F \right] \left[l^1 \right] \left[l^{-3} \right] \left[l^{4-F} \right]^{1/2}}$$

(3.5-d)

where x represents distance, and V represents volume. From Equation (3.4) it follows that:

$$F = \begin{bmatrix} l^1 \\ l^2 \\ l^3 \\ l^4 \end{bmatrix} \Rightarrow a = \begin{bmatrix} l^{-2} \\ l^{-1} \\ l^0 \\ l^1 \end{bmatrix} \Rightarrow \tau = \begin{bmatrix} l^{1.5} \\ l^1 \\ l^{0.5} \\ l^0 \end{bmatrix}$$

$$\Rightarrow P = \begin{bmatrix} l^{-2.5} \\ l^{-1} \\ l^{0.5} \\ l^2 \end{bmatrix} \Rightarrow \frac{P}{V} = \begin{bmatrix} l^{0.5} \\ l^2 \\ l^{3.5} \\ l^5 \end{bmatrix}$$

(3.6)

3.2   Non-dimensional parameters

Most phenomena in fluid mechanics and heat transfer depend on geometric and flow parameters in a complicated way. The solution to this difficulty lies in dynamical similarity such that complex physical phenomena may be characterized using only a few dimensionless groups.

Forces encountered in flowing fluids include those due to inertia, viscosity, pressure, gravity, surface tension, and compressibility (or density gradients). A dimensionless number representing the relative magnitude of the inertia to viscous effects in a fluid flow, liquid or gas, is the Reynolds number defined as:

$$Re = \frac{\rho U L}{\mu} \tag{3.7}$$

where U and L are the flow characteristic velocity and length scale, while ρ and μ are the fluid density and viscosity, respectively. The Reynolds number has been the control parameter distinguishing between creeping ($Re \rightarrow 1$), viscous flow ($Re > 1$) and inviscid flow ($Re \rightarrow \infty$). The Reynolds number is also the control parameter characterizing laminar and turbulent flows either in internal, e.g. channel, or external flows, e.g. boundary layers.

In the analysis of similarities between fluid flow and heat transfer, the important control parameter is the Prandtl number, defined as:

$$Pr = \frac{c_p \mu}{k_f} \tag{3.8}$$

where k_f and c_p are the fluid thermal conductivity and specific heat, respectively. Prandtl number is a thermophysical property expressing the ratio between the momentum diffusion through the fluid, due to viscosity, and the heat diffusion by conduction. Gases such as air have Prandtl number close to unity and, hence, the velocity and thermal boundary layers are of similar thickness.

In some instances, the continuity and momentum equations can be decoupled from the energy equation. Then, the velocity field is independent of the temperature distribution and can be determined, regardless of the heat transfer conditions imposed on the flow. The solution for the velocity field can be substituted into the energy equation to determine the temperature distribution. The resulting temperature field then depends on the Peclet number, which is a product of Reynolds and Prandtl numbers, defined as:

$$Pe = \frac{UL}{\alpha} = Re \cdot Pr \qquad\qquad (3.9)$$

where $\alpha = k_f / \rho c_p$ is the thermal diffusivity. Peclet number is a measure of the relative magnitude of heat transfer by convection to heat transfer by conduction. Since the Reynolds number is required independently for dynamical similarity, it is customary to work with the Reynolds and Prandtl numbers separately rather than the Peclet number.

The primary parameter of interest in convective heat transfer is the heat transfer coefficient, h, defined by Newton's law of cooling (Equation 2.9). The most convenient dimensionless form of the heat transfer coefficient is the Nusselt number defined as:

$$Nu = \frac{hL}{k_f} \qquad\qquad (3.10)$$

which is the ratio of convection heat transfer to fluid conduction heat transfer under the same conditions. A Nusselt number of order unity would indicate a sluggish motion as effective as pure fluid conduction, e.g. laminar flow in a long pipe. A large Nusselt number means very efficient convection, e.g. turbulent pipe flow with Nu in the range 100-1000. By far, the main effort in convective heat transfer problems is dedicated to forming the appropriate correlation between the Nusselt number and the Reynolds and Pradtl numbers.

The Nusselt number is not the only way to nondimensionalize the heat transfer coefficient. A widely used alternative parameter is called the Stanton number defined as:

$$St = \frac{h}{\rho U c_p} = \frac{Nu}{Re \cdot Pr} \qquad\qquad (3.11)$$

Indeed, the principles of dimensional analysis admit alternative parameters that are new but not independent. The Stanton number is simply regrouping of the Reynolds, Prandtl, and Nusselt numbers.

A prime parameter affecting heat conduction in solids with convection boundary conditions is the Biot number, which is identical in form to the Nusselt number used in heat convection. However, for Biot number, the heat convection is normalized by the thermal conductivity of the solid k_s rather then that of the fluid, i.e.:

$$Bi = \frac{hL}{k_s} \qquad (3.12)$$

This parameter represents the ratio between the internal conduction resistance and the external convection resistance to the flow of heat. In unsteady heat transfer problems, the lumped system analysis can be used if the Biot number is small. For $Bi<0.1$, the temperature distribution during transients within the solid at any instance is assumed to be uniform, which is a great simplification.

In convective heat transfer, a portion of the mechanical energy is converted by shear stresses into thermal energy, increasing the temperature of the working fluid even if there is no heat transfer boundary condition (adiabatic wall). The Eckert number is a measure of the relative importance of viscous dissipation given by,

$$Ec = \frac{U^2}{c_p \Delta T} \qquad (3.13)$$

where ΔT is the increased fluid temperature due to friction. The viscous dissipation is important in high-speed flows, such as in aerodynamics, or highly viscous fluids at modest velocities, such as lubricating oils.

Another parameter governing the temperature increase due to viscous dissipation is the Mach number,

$$Ma = \frac{U}{a} \qquad (3.14)$$

where a is the speed of sound. The Mach number is of course used to categorize gas flows into subsonic and supersonic flows. However, it is also used to indicate when dissipation and compressibility effects are important. Flows with $Ma<0.1$ are considered incompressible.

The ratio between the viscous dissipation and fluid conduction effects is represented by the Brinkman number defined as:

$$Br = \frac{\mu U^2}{k_f \Delta T} \qquad (3.15)$$

For low-speed flows, only the most viscous fluids have significant Brinkman number.

The Grashof number is a measure of the relative size of buoyant forces, due to heating, to viscous forces defined as,

$$Gr = \frac{g\beta L^3 \Delta T}{v^2} \qquad (3.16)$$

where β is the coefficient of the fluid volume expansion, and $v=\mu/\rho$ is the fluid kinemtaic viscosity. When the buoyant force is negligible compared to the viscous or inertia forces, i.e. very small Grashof number, pure forced convection is expected. If the buoyant force is dominant, very high Grashof number, pure free convection is expected. Thus, the heat transfer correlation for pure free convection takes the form: $Nu=f(Gr,Pr)$, where the Grashof number performs the same role of the Reynolds number in pure forced convection, $Nu=f(Re,Pr)$. However, for this purpose, the Rayleigh number:

$$Ra = \frac{g\beta L^3 \Delta T}{v^2} \frac{\mu c_p}{k_f} = GrPr \qquad (3.17)$$

a product of the Grashof and Prandtl numbers, is used more often.

Heat transfer with phase change, in either a boiling or a condensation process, introduces additional parameters. The Jakob number is a measure of the relative surface-saturation temperature excess defined as,

$$Ja = \frac{c_p \Delta T}{h_{fg}} \qquad (3.18)$$

where h_{fg} is the latent heat. Convection terms are not negligible for $Ja \gg 1$, where the wall temperature is either very cold in a condensation or very hot in a boiling process.

In two-phase flows, with or without phase change, interfaces between vapour or gas phase and liquid phase exist, giving rise to surface tension effects. The Bond number is the ratio of gravity force to surface tension force and arises in any case where gravity and surface tension interact.

$$Bo = \frac{g\Delta\rho L^2}{\sigma} \qquad (3.19)$$

Gravity is a body force, which is expected to be very small on a microscale. Indeed, this effect is most readily demonstrated by the rise of liquids against gravity in a small-diameter capillary.

Surface tension effect is characterized by the Weber number defined as,

$$We = \frac{\rho L U^2}{\sigma} \tag{3.20}$$

The Weber number is typically very large, since surface tension is small, and hence has negligible effect unless the characteristic scale or the surface curvature is very small. It follows that surface tension is important only for low velocities in physically small free-surface flows.

Transient heat transfer is characterized by variations in time t, and a dimensionless time is expressed by the Fourier number

$$Fo = \frac{\alpha t}{L^2} \tag{3.21}$$

This definition of the Fourier number indicates that bodies with a high diffusivity respond faster than those with a low diffusivity, while large bodies respond slower than small bodies.

3.3   Similarity parameters in convective heat transfer

The power of dimensional analysis is to simplify the characterization of physical problems. However, for most practical problems, even a reduced number of dimensionless parameters is usually too large to allow simple numerical and analytical modelling. The solution to this difficulty lies in approximation and simplification, which can be done either intuitively or by exploring asymptotic limits.

In convective heat transfer, the heat transfer coefficient depends on many parameters. For single-phase convection, this dependence can functionally be expressed in terms of fluid properties, flow parameters and boundary conditions, as follows [218]:

$$h = h\left(\rho, \mu, k, c_p, \beta, \sigma; U, P, T, L; g, q''; \text{shape}\right) \tag{3.22}$$

or

$$Nu = Nu\left(Re, Pr, Gr, Ec; \text{shape}\right) \tag{3.23}$$

The implications of Equation (3.22) are twofold. It predicts the form in which an analytical solution to a convective heat transfer problem will be expressed. It also suggests the proper correlation of experimental data reducing the number of variables to four only.

In two-phase flows with phase change, the heat transfer coefficient could further depend on the difference between the surface and saturation temperatures, $\Delta T = |T_{sur} - T_{sat}|$, the body force arising from the liquid-vapour density difference, $g(\rho_l - \rho_v)$, the latent heat, h_{fg}, the surface tension, σ, a characteristic length scale, L, and the thermophysical properties of the liquid or vapour, ρ, c_p, k, μ. That is:

$$h = h\left[\Delta T, g\left(\rho_l - \rho_v\right), h_{fg}, \sigma, \rho, \mu, k, c_p, L, U; \text{shape}\right] \quad (3.24)$$

Invoking the Buckingham Pi-theorem, five groups are expected. The groups can either be expressed in the following forms:

$$\frac{hL}{k} = f\left[\frac{\rho g\left(\rho_l - \rho_v\right)L^3}{\mu^2}, \frac{c_p \Delta T}{h_{fg}}, \frac{\mu c_p}{k}, \frac{g\left(\rho_l - \rho_v\right)L^2}{\sigma}, \frac{\rho U^2 L}{\sigma}\right] \quad (3.25)$$

or identifying the dimensionless groups as:

$$Nu = f\left[\frac{\rho g\left(\rho_l - \rho_v\right)L^3}{\mu^2}, Ja, Pr, Bo, We\right] \quad (3.26)$$

The Nusselt and Pradtl numbers are familiar from single-phase convection analysis, while the Bond and Weber numbers have already been discussed. The new dimensionless parameters are the Jakob number Ja, the Bond number Bo, and a parameter that bears a strong resemblance to the Grashof number. This parameter represents the effect of buoyancy-induced fluid motion on heat transfer. The Jakob number is the ratio of the maximum sensible energy absorbed by the vapour (liquid) to the latent energy absorbed by the vapour (liquid) during boiling (condensation). In many applications, the sensible energy is much less than the latent energy, and Ja has a small numerical value.

3.3.1   Dimensional analysis for microscale heat convection

A motivating exercise to show the advantages of utilizing single-phase flow in smaller channel size, using classical correlations, has been presented recently by Palm [140]. In a fully-developed laminar flow, the Nusselt number is constant. This means that the heat transfer coefficient h is inversely proportional to the hydraulic diameter D_h, defined as four times the cross-sectional area divided by the wetted perimeter, such that:

$$Nu = \frac{h \cdot D_h}{k_f} = const \quad \Rightarrow \quad h \propto \frac{1}{D_h} \qquad (3.27)$$

where k_f is the fluid thermal conductivity. In order to transfer a certain amount of heat q at a given mean temperature difference ΔT, the product of the surface area A and the heat transfer coefficient must be constant. Hence, considering n parallel channels, the product of the channel length L and the number of channel should be constant,

$$q = h \cdot A \cdot \Delta T = h(\pi \cdot D_h \cdot L \cdot n)\Delta T \quad \Rightarrow \quad L \cdot n = const \qquad (3.28)$$

The friction factor in fully developed laminar flow is inversely proportional to the Reynolds number, Re, and the pressure drop is inversely proportional to the quadratic power of the hydraulic diameter (classical Poiseuille flow):

$$\Delta P = \frac{C}{Re} \rho u^2 \frac{L}{D_h} = \frac{Cv}{uD_h} \rho u \left(\frac{Q_v}{\pi D_h^2 / 4} \right) \frac{L}{D_h} \quad ; \quad \Delta P \propto \frac{1}{D_h^4} \qquad (3.29)$$

This shows that for a given volume flow rate, Q_v, under a certain pressure drop, ΔP, specific ratio between the channel size and the number of channels must be kept constant as follows:

$$\Delta P = \frac{C'v \cdot \rho \cdot L}{D_h^4} \left(\frac{Q_v}{n} \right) \quad \Rightarrow \quad \frac{L}{n \cdot D_h^4} = const \qquad (3.30)$$

However, since the product $(L \cdot n)$ must be constant to maintain the mean temperature difference, the product $(n \cdot D_h^2)$ should also be constant.

In summary, to maintain the pressure drop and the mean temperature difference, at a given mass and heat flow rate while changing the channel size and the number of channels, the product $n \cdot D_h^2$ should be kept constant. The channel length is then given by the condition: $L \cdot n$=const. Applying these simple relations leads to the results illustrated in Figure 3.3. As an example, decreasing the diameter by half would double the heat transfer coefficient and reduce the necessary surface area to one half of the original. The channel length would be reduced to one fourth of the original, the number of parallel channels would increase by a factor of 4, and the internal volume, $nL\pi D_h^2/4$, would decrease by a factor of 4. In addition, the channel wall thickness required to withstand a certain pressure would decrease. Thus, a decrease of the channel size leads to a much more compact design.

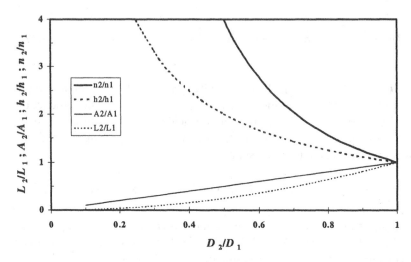

Figure 3.3: Dependence of duct length, number of parallel ducts, heat transfer surface area, and heat transfer coefficient on the duct hydraulic diameter for given mass and heat flow rate, mean temperature difference and total pressure drop.

3.4 Control parameters in microscale convective heat transfer

Three size effects have been identified in Chapter 2, slip effect for gas and electrokinetic and polarity effects for liquid flow in microducts. Each of these physical mechanisms is associated with an inherent length scale. Naturally, the ratio between each of these length scales and the external flow scale, i.e. the duct size, will dictate the impact of the size effect on the evolution of fluid flow and heat transfer. Clearly, as the flow scale increases beyond a certain value, depending on the particular physical mechanism, the size effect can be neglected such that classical theories for macrosystems are valid. It is also important to keep in mind that when the flow scale decreases beyond a certain value, the corrections due to the size effects may not be applicable and a different approach would be required.

3.4.1 Knudsen number

The physical model of gas as a collection of molecules of a finite size, executing a random motion with near perfect intermolecular collisions, gives rise to two inherent length scales: (i) average distance between molecules, and (ii) average distance between successive molecular collisions. Transport phenomena such as diffusion, heat conduction and viscosity depend directly on molecular collisions. Therefore, the latter length scale, called the mean free path λ, is the important parameter in any phenomenon depending on the transport of molecules [91]. The mean free path depends on the gas pressure and temperature as follows:

$$\lambda = \frac{\mu}{P} \left(\frac{\pi RT}{2} \right)^{1/2} \tag{3.31}$$

where R is the specific gas constant. The ratio between the mean free path and the flow characteristic scale is the celebrated Knudsen number,

$$Kn = \frac{\lambda}{L} \tag{3.32}$$

The assumption of $Kn \to 0$ has been the cornerstone assumption for the development of continuum-based theoretical models for fluids. Indeed, in most common gas systems, the mean free path is much smaller than any flow scale. However, in cases where the mean free path is very large, such as in re-entry of space vehicles problems, or the flow scale is very small, such as in microchannels, the Knudsen number is no longer small. A classification of flow regimes based on the knudsen has been introduced in Chapter 2.

3.4.2   Electrokinetic parameter

Electrostatic force is present between charged molecules or particles. The force has an inverse-square dependence on the distance. Any solid surface is likely to carry some charge because of broken bonds and surface trap charge. If the surface is a good insulator, trapped charges can induce very high voltage [69]. Charged surfaces in liquids can have an important effect on the fluid flow and heat transfer due to the charge re-distribution in the liquid. The characteristic thickness of the electric double layer (EDL), in which the electrostatic potential decays exponentially, is the Debye length given by:

$$\lambda_D = \left(\frac{\varepsilon \varepsilon_0 k_b T}{2 n_0 z^2 e^2} \right)^{1/2} \tag{3.33}$$

where ε is the dielectric constant of the medium, ε_0 is the permittivity of vacuum, and k_b is the Boltzmann constant. z is the valence of negative and positive ions, e is the electron charge, and n_0 is the ionic concentration. For example, the Debye length in pure water is about $1\mu m$, while it is only 0.3nm in 1 mole of NaCl solution. The EDL effect on the flow field depends on the ratio between the Debye length and the flow scale, which can be called the Debye number defined as:

$$De = \frac{L}{\lambda_D} \qquad (3.34)$$

The EDL effect can be neglected if the flow scale is large or the Debye scale is very small, i.e. $De \gg 1$, such as in concentrated ionic solutions. However, the effect can be significant if the flow scale is very small, i.e. $De \sim 1$.

3.4.3   Polarity parameter

In micro polar fluid, the inherent length scale arises from the ratio among different viscosity coefficients involved.

$$l = \left[\frac{\gamma}{\mu} \left(\frac{\mu + \kappa}{\kappa} \right) \right]^{1/2} \qquad (3.35)$$

When viscous effects are much larger than the couple stress effects, $\gamma/\mu \to 0$, this length scale vanishes $l \to 0$. On the other hand $l \to \infty$ either (i) when the couple stresses effects are much larger than the viscous effects, i.e. $\gamma/\mu \to \infty$, or (ii) when the the effect of microrotation is negligible in comparison to the viscous effect, i.e. $\kappa/\mu \to 0$. Since polar mechanics in general and the theory for micropolar flow in particular originated by Eringen [40], it may be appropriate to call the ratio between the polarity length scale and the flow scale the Eringen number defined as:

$$Er = \frac{L}{l} \qquad (3.36)$$

Similar to the Knudsen number, as $Er \to \infty$ micropolar flow effect becomes negligible and the classical non-polar mechanics is valid. However, as $Er \to 1$ micropolar flow effects have to be considered.

Chapter 4

Fabrication of Thermal Microsystems

Fabrication of thermal microsystems involves techniques that were initially developed for the fabrication of integrated circuits (ICs) and later extended to the fabrication of microelctromechanical systems (MEMS). Analytical and experimental aspects of these techniques have been presented in great detail in many books and, therefore, will not be discussed here. However, the fabrication of thermal microsystem for either commercial applications or microscale heat transfer research imposes contradicting demands. This challenge can be met by utilizing standard microfabrication technologies in a unique and innovative approach. Therefore, only processes and techniques developed and used specifically for the fabrication of the two case studies, i.e. microchannel heat sinks and micro heat pipes, will be described hereafter.

4.1   Fabrication of an integrated microchannel heat sink

A microchannel heat sink integrated with a local heater, serving as the heat source, and microsensors for temperature measurements can be fabricated in a CMOS-compatible process. Hence, such a microsystem can be further integrated with ICs or other micro components. Complete analysis of the thermal microsystem performance would require information about the evolving flow patterns during operation. Hence, an optically clear path into the flow cell is needed, which can be furnished by using a transparent cover. Indeed, the process can be modified to include glass as the cover material; thus, enabling flow visualizations. However, the inclusion of glass renders the process non CMOS-compatible, due to cross-contamination concerns. Therefore, both the CMOS- and the non CMOS-compatible fabrication processes of the thermal microsystem will be presented.

4.1.1   CMOS-compatible fabrication process

The novel idea of maskless and self-aligned silicon etch between bonded wafers is a result of the orientation dependent and self-stopping silicon anisotropic etch in (100) wafers [82]. One wafer is first slightly etched using a patterned oxide as the etch mask. After stripping the oxide, a second wafer is bonded to the patterned wafer resulting in shallow gaps at the desired locations. Etching holes are then drilled through either one of the wafers to allow the supply of the etchant into the gaps. Anisotropic etching of both wafers proceeds until it stops on the (111) surfaces. Thus, the silicon etch of both wafers can be accomplished such that it is self-aligned and self-stopped.

A variety of thermal microsystems consisting of numerous microchannels, a localized heater and an array of temperature micro-sensors have been fabricated on a single die, about 10mm×20mm in area. Schematic cross sections of the major fabrication steps are shown in Figure 4.1. The starting substrates are a pair of 100mm in diameter, (100) orientated, p-type silicon wafers with thickness of 525μm. The fabrication begins with thermal growth of a 0.3μm oxide layer on both wafers. On the front side of the device wafer, a 0.3μm LPCVD 840°C low stress silicon nitride is deposited for insulation. A 0.4μm polysilicon film is next deposited on the nitride and patterned to serve as the heater and sensor material. The dumbbell-shaped thermistors are formed by selectively implanting phosphorus into a 8μm stretch of polysilicon in the center of the dumbbell, with 5×10^{13}/cm^2 dosage and 160KeV implant energy. The sensor leads and the serpentine-shaped heater are heavily doped with 5×10^{15}/cm^2 arsenic to reduce the resistivity. The doped polysilicon is then annealed at 900°C for 30min.

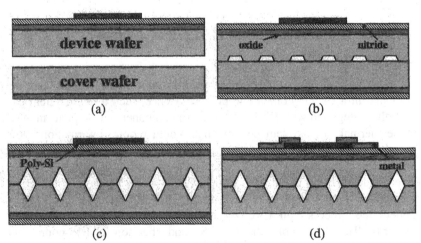

Figure 4.1: Schematic cross-sections of the microchannel heat sink major fabrication steps using CMOS-compatible technology.

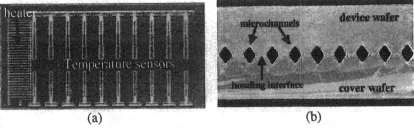

(a)   (b)

Figure 4.2: Potographs of (a) a die front side showing the heater, temperature microsensors and metal lines, and (b) a SEM cross-section showing the diamond-shaped microchannels between the bonded wafers.

The oxide layer on the backside of the device wafer is patterned and 5μm-deep grooves are etched using a 25wt% aqueous TMAH solution at 85°C. This is followed by stripping the oxide etch mask from the device wafer as well as the oxide layer on the bonding surface of the cover wafer (Figure 4.1(a)). The exposed silicon surfaces of both wafers are boron doped at 1050°C for 1hr to improve the fusion bonding quality. Following vacuum pre-bonding of the two wafers at room temperature, the wafers are then annealed at 1050°C for 1hr (Figure 4.1(b)). Next, fluid inlet/outlet holes, 750μm×750μm in area, are etched through the cover wafer into the grooves using 25wt% aqueous TMAH solution at 85°C. The solution flows into the grooves and completes the etching of the channels in both wafers (Figure 4.1(c)). Finally, a 1μm Al-Si layer is sputtered, patterned, and sintered at 450°C for 30min to form the interconnects (Figure 4.1(d)). A photograph of a fabricated device is shown in Figure 4.2. The local heater, temperature sensors and metal lines can be seen in the picture of the device front side (Figure 4.2(a)). The cross-sectional SEM picture demonstrates the diamond-shaped, self-aligned and self-stopped microchannels (Figure 4.2(b)).

4.1.2   Glass-based fabrication process

Each device is designed as an integrated microsystem consisting of an array of microchannels (30 to 40 in number), a localized heater and a 2-D array of temperature microsensors on a single die, 10mm×20mm in area [85]. The micro heat sink includes an array of grooves, formed by bulk Si etch in TMAH, either 50μm or 100μm in nominal width and about 18mm in length. A Pyrex 7740 glass wafer is anodically bonded to the silicon substrate to cap the grooves. The resulting transparent microchannels, with triangular cross-sections, allow in-situ video recording of the flow pattern during device operation. The integrated heater and distributed temperature microsensors are fabricated on the silicon substrate using standard surface micromachining technology.

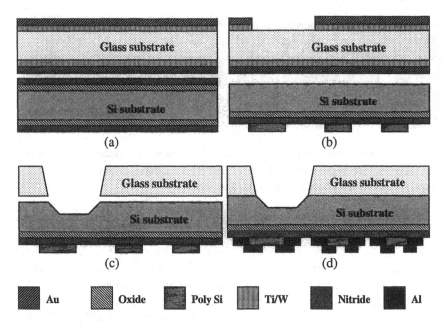

(a)   (b)

(c)   (d)

| ▨ Au | ▨ Oxide | ▨ Poly Si | ▨ Ti/W | ▨ Nitride | ▨ Al |

Figure 4.3: Schematic cross-sections of the microchannel heat sink major fabrication steps using glass-based technology.

Schematic cross sections of the major fabrication steps are shown in Figure 4.3. The starting substrate is a (100) orientated, p-type silicon wafer 100mm in diameter, 525μm in thickness, and resistivity of about 17Ω·cm. Similar to the previous microsystem, the fabrication begins with thermal growth of a 0.3μm oxide layer, followed by LPCVD of a 0.3μm low stress silicon nitride layer at 840°C for insulation (Figure 4.3(a)). A 0.4μm polysilicon film is deposited on top of the nitride and patterned to serve as the heater and sensor material (Figure 4.3(b)). The dumbbell-shaped thermistors are formed by selectively implanting phosphorus into an 8μm stretch of polysilicon in the center of the dumbbell, with 5×10^{14}/cm^2 dosage and 60keV implant energy. The remaining portions of the sensors and the serpentine-shaped heater are heavily doped with arsenic, dosage of 8×10^{15}/cm^2, to reduce the resistivity. The doped polysilicon is then annealed at 900°C for 30min. Next, the oxide layer on the wafer backside is patterned to serve as an etch mask, followed by bulk Si etching in a 25wt% aqueous TMAH solution at 85°C to form the microchannel array (Figure 4.3(c)). After the oxide etch mask is stripped, a 1μm Al-Si layer is sputtered on the front side of the wafer, patterned and sintered at 450°C for 30 minutes to form the interconnects (Figure 4.3(d)).

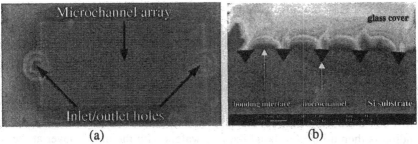

(a)   (b)

Figure 4.4: Potographs of (a) a die backside showing the array of microchannels and both inlet/outlet holes, and (b) a SEM cross-section showing the traingular-shaped microchannels between the bonded wafers.

In parallel, a stack of a 0.4μm Au layer on top of a 0.04μm Cr thin film is sputtered on a Pyrex 7740 glass wafer to serve as an etch mask (Figure 4.3(a)). After patterning the holes for the channel inlet/outlet (Figure 4.3(b)), the glass wafer is etched through in 40% HF, followed by stripping the Cr/Au layers (Figure 4.3(c)). The processed silicon and glass wafers are then aligned and pre-bonded together, immediately after a cleaning process, and are anodically bonded at $340°C$ with applied voltage of 700V (Figure 4.3(d)). Pictures of a fabricated device are shown in Figure 4.4. The transparent microchannel array in the silicon and the inlet/outlet holes in the glass wafer are clearly visible in the die backside picture shown in Figure 4.4(a). The die front side is identical to the CMOS-compatible microsystem shown in Figure 4.2(a). The device cross-sectional SEM picture in Figure 4.4(b) illustrates the triangular shape of the microchannels buried between the bonded silicon and glass wafers.

4.2 Fabrication of an integrated micro heat pipe

Miniature heat pipes are currently used as a part of cooling systems for high-performance microprocessors in laptop computers and notebooks. However, these mini heat pipes are fabricated separately, using traditional manufacturing tools, and assembled at a later stage. These cooling systems are not integrated with the ICs, in a batch fabrication process, partly due to the lack of compatible technology. The fabrication of a complete micro heat pipe system in a fully compatible CMOS process is described. Again, a modified approach is presented for capping the micro heat pipes with glass, in a non CMOS-compatible process, to allow in-situ flow visualizations. Furthermore, in two-phase flows such as the micro heat pipe flow, the vapour and liquid content in the mixture is an important parameter. Thus, a unique attempt to integrate capacitive sensors with a micro heat pipe system to locally measure the void fraction is also discussed.

4.2.1   CMOS-compatible fabrication process

An integrated microsystem comprising a microchannel array, a local heater, temperature and capacitance microsensors is designed to operate as a micro heat pipe system. In a fully CMOS compatible process, triangular grooves are etched in a Si wafer using TMAH, and the grooves are capped by a thin transparent film using wafer bond and etch back technology [109]. A low-stress silicon nitride film is deposited on another Si wafer, and fusion bonding is then utilized to bond the two wafers with the nitride layer at the interface. The handling wafer is then dissolved in TMAH, while the device wafer with the grooves is protected. The etching process stops on the nitride film, resulting in grooves capped by the nitride thin film. This allows the integration of other microelectronic devices using CMOS technology on top of the processed wafers.

Two electrodes are required for a capacitance sensor. The bottom electrodes for the capacitors are created by doping the grooves prior to bonding, while metal pads are formed on the nitride membrane after bonding as top electrodes. The capacitance between the metal pads and the doped substrate is a function of the dielectric constant of the media between the two electrodes. Hence, each sensor can be used for measuring the local void fraction, on which the local fluid flow dielectric constant depends.

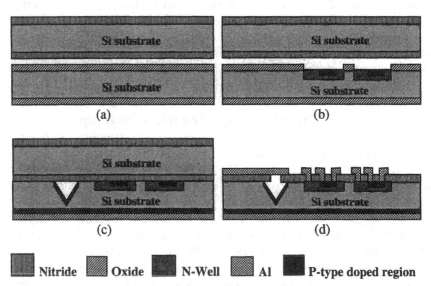

Figure 4.5: Schematic cross-sections of the micro heat pipe major fabrication steps using CMOS-compatible technology.

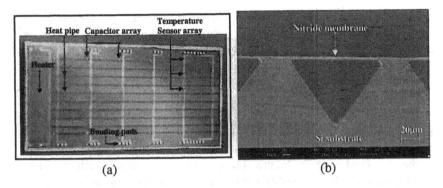

Figure 4.6: Potographs of (a) a die front side showing the heater, temperature and capacitance microsensors, micro heat pipe array and metal lines with bonding pads, and (b) a SEM cross-section showing the triangular-shaped micro heat pipe capped by a thin and flat silicon nitride membrane.

The heater, placed at one edge of the device to serve as a heat source, is formed in the silicon substrate to minimize the heat loss to the surrounding. Both the heater and the temperature microsensors are fabricated by selective boron implantation in n-wells, which are defined by phosphorus implantation. The temperature microsensors are located along the micro heat pipes to measure the temperature distribution.

Schematic cross-sections of the main fabrication steps are drawn in Figure 4.5. The fabrication of the micro heat pipe starts with 0.3μm thermal oxide growth on a (100), p-type Si wafer with 100mm in diameter and 525μm in thickness (Figure 4.5(a)). N-Wells are formed by selective phosphorus implantation, with dosage of 1×10^{13}/cm^2 and energy of 120keV, followed by activation and drive-in. Boron is selectively implanted within the n-wells to create the heater and the temperature microsensors (Figure 4.5(b)). The sensing elements, at the centre of each dumbbell-shaped temperature sensor, are lightly doped (dosage of 1×10^{14}/cm^2 and energy of 30keV), while the heater and sensor leads are moderately doped with boron (dosage of 5×10^{15}/cm^2 and energy of 30keV). After deposition and densification of a 0.5μm-thick LTO layer, V-grooves are etched using 25wt% aqueous TMAH solution at 85°C to form the triangular-shaped pipes. Boron is diffused into the grooves and the wafer backside to serve as the capacitors bottom electrode (Figure 4.5(c)). Simultaneously, a 0.5μm-thick nitride layer is deposited on the cover Si wafer (Figure 4.5(a)). The cover and device wafers are bonded together by fusion bonding at 1050°C in O$_2$ ambient. A 0.5μm-thick LTO layer is deposited and densified to protect the device wafer backside (Figure 4.5(c)). The cover wafer is then dissolved in TMAH solution exposing the etch-stop nitride layer, which covers the V-grooves. Contact windows are next opened in the nitride layer for the heater and temperature sensors, followed by sputtering and patterning a 1μm-thick aluminium layer to form the interconnects and capacitor top electrodes. Finally, the liquid charging holes, one at the end of each micro heat pipe, are opened in the nitride membrane using dry etch (Figure 4.5(d)).

Photographs of a fabricated micro heat pipe system, using fully CMOS-compatible technology, are shown in Figure 4.6. The overall area of the die is 10mm×20mm. The heater region, micro heat pipe array, temperature microsensors, capacitor electrodes and metal lines can be seen in the picture of the die front side in Figure 4.6(a). The device cross-sectional SEM picture, shown in Figure 4.6(b), demonstrates the success of the wafer bond and etch back technology in capping the triangular-shaped heat pipes with a thin nitride film, which survived the rest of the fabrication process.

4.2.2   Glass-based fabrication process

The transparency of the nitride film is not as good as glass. Therefore, in order to enhance the quality of flow visualizations, a glass wafer is designed to cover the grooves [109]. An aluminium layer is first sputtered and patterned on a glass wafer for both metal interconnects and capacitor top electrodes. Next, holes are drilled through the glass wafer, two holes per groove, to facilitate charging of the heat pipes with working liquid. The glass wafer is then anodically bonded to a silicon wafer, with etched grooves, to provide an optically clear cover.

Each die contains four triangular micro heat pipes, etched in the Si wafer, with an overall area of 15mm×30mm. The heat pipes are 100μm in width, at the base, and 20mm in length. Both the heater and temperature sensors are fabricated in the Si wafer using selective boron implantation within n-wells, which are formed by phosphorus implantation. The heater is located at one edge of the Si die, serving as the heat source, while the temperature sensors are located along the heat pipes. The other edge of the Si die is designed to allow an intimate contact to a cold heat sink for heat dissipation. The common bottom electrode of all capacitors is defined by doping the Si grooves before bonding. The capacitor top metal electrodes are fabricated on the glass bonding-surface. In order to minimize the parasitic capacitance of the sensors, all the pads for wirebonds are fabricated on the glass wafer. Hence, the metal patterns on the Si and glass wafers are designed to cross each other and maintain intimate contact after the anodic bonding. This allows the transmission of all electronic signals through the metal lines on the glass. The glass die is wider than the silicon die to minimize the overlap are between the bonding pads and the silicon substrate. It is critical to ensure the integrity of the metal lines through the interface between the glass and the silicon edge during bonding to avoid cuts in the metal lines. Since a large portion of the capacitor leads and all wirebond pads are fabricated on the glass substrate, the overlap area between the Al metal and the silicon substrate is minimized. The designed overlap area of the metal lines is about $8500\mu m^2$ per sensor. The parasitic capacitance is estimated to be about 1pF if a $0.3\mu m$-thick oxide layer is used for electric insulation between the metal lines and the silicon substrate.

Schematic cross-sections of the major fabrication steps are shown in Figure 4.7. A $0.3\mu m$-thick thermal oxide layer is firstly grown on a (100), p-type silicon wafer (Figure 4.7(a)). N-wells are patterned and formed by phosphorus implantation, followed by activation and drive-in. Boron is selectively implanted within the n-wells to construct the heater and the temperature microsensors (Figure 4.7(b)). The formation of the n-wells, heaters and temperature sensors is identical to the CMOS-compatible process described above. After the deposition and patterning of a $0.5\mu m$-thick LTO layer, V-grooves are etched in the silicon substrate using TMAH to create the triangular-shaped heat pipes. Boron is diffused into the grooves and the wafer backside to form the capacitor bottom electrode. A $0.1\mu m$-thick oxide, thermally grown in the process, is used for electrical insulation (Figure 4.7(c)). Contact windows are then opened for the heater and temperature sensors, followed by sputtering and patterning of a $0.15\mu m$ aluminium layer for the interconnections (Figure 4.7(d)).

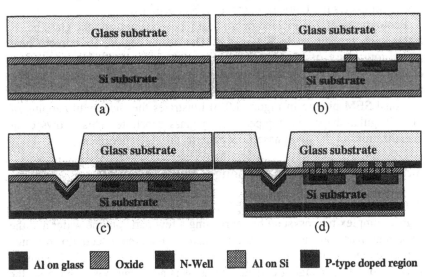

Figure 4.7: Schematic cross-sections of the micro heat pipe major fabrication steps using glass-based technology.

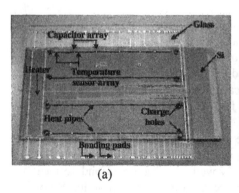
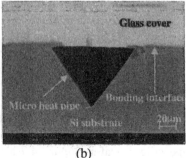

(a)   (b)

Figure 4.8: Potographs of (a) a die front side showing the heater, temperature and capacitance microsensors, metal lines with the bonding pads on the galss and four micro heat pipes, each with two charging holes, and (b) a SEM cross-section showing the triangular-shaped micro heat pipe capped by an anodically bonded glass substrate.

In parallel, a 0.15μm-thick Al layer is sputtered and patterned on a Pyrex 7740 glass wafer for both the metal interconnects and the capacitor top electrodes (Figure 4.7(b)). Then two liquid charging holes for each heat pipe are drilled through the glass (Figure 4.7(c)). Finally, die-by-die anodic bonding of the silicon and glass substrates, at 330°C under applied voltage of 700V, completes the fabrication process (Figure 4.7(d)).

Photographs of a fabricated micro heat pipe system, using the glass-based technology, are shown in Figure 4.8. The heater region, four micro heat pipes, four pairs of charging holes, temperature microsensors, capacitor electrodes and metal lines with bonding pads on the glass substrate can be seen in the picture of the die front side in Figure 4.8(a). The device cross-sectional SEM picture in Figure 4.8(b) illustrates the successful capping of the triangular-shaped heat pipes with a glass substrate. This allows clear visualizations of the two-phase flow patterns during experiment.

4.3   Technological challenges and solutions

One of the advantages of using wafer bonding is the option to realize a highly complex microsystem by fabricating a few parts on one wafer and the remaining parts on the other. It is not hard to imagine a scenario in which certain elements on each wafer require metal interconnects for I/O signals. Clearly, due to packaging considerations, it is advantageous to have all the wire bonds on the same side of a die. Moreover, it is beneficial to form the bonding pads on a glass wafer, whenever possible, due to its superior dielectric properties.

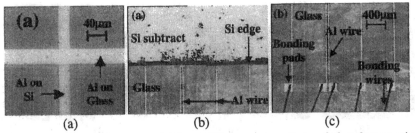

Figure 4.9: Close-up pictures of: (a) metal lines on opposite bonded wafers crossing each other, (b) metal lines observed from the Si substrate side crossing the Si die edge, and (c) metal lines ending at bonding pads on the glass side.

The picture in Figure 4.9-a illustrates a cross-connection between metal lines on the two substrates resulting from the silicon-to-glass anodic bonding. Thus, all the I/O signals could be transmitted through the lines on the glass. The smaller silicon die allowed the formation of the bonding pads on the glass substrate, as shown in Figure 4.9-b, to eliminate the overlap between the metal pads and the silicon substrate. Consequently, all electronic signals are channelled through the metal lines to the bonding pads on the glass resulting in a significant reduction of the parasitic capacitance. The continuity of the metal lines ending at bonding pads on the larger glass die is demonstrated in Figure 4.9-c.

A variety of problems could arise during the fabrication and operation of glass based integrated microsystem [109]. For example, a micro heat pipe system was fabricated using a similar die-by-die bonding technique. The edges of the individual silicon dies were jagged due to the wafer die sawing. Hence, the metal lines on the glass crossing the edge of the Si die were often broken during bonding, as demonstrated in Figure 4.10-a, resulting in open circuits. Moreover, the capacitor top electrodes made of aluminum were in direct contact with water inside the heat pipes. The Al electrodes were consequently oxidized very rapidly, as shown in Figure 4.10-b, such that the metal electrodes no longer functioned. Finally, the anodic bonding between the glass and silicon substrates requires flat bonding surfaces to achieve good bonding quality with high yield. Therefore, the metal lines forming the heater and temperature sensors were very thin, about 0.15µm, in order to minimize topography variations on the bonding surfaces. The Al thin film forming sensors and interconnects was adequate since low current was required for the operation of these elements. However, the heater required high current to pass through in order to provide adequate heating of the device. As a result, the thin Al wire could not sustain the high current and burned out during operation, as shown in Figure 4.10-c. Hence, the fabrication process of the integrated micro heat pipe system has to be modified in order to overcome these difficulties.

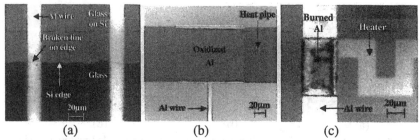

Figure 4.10: Pictures of (a) Al wires broken during bonding by the rough Si die edge, (b) an oxidized capacitor top electrode due to direct contact with the working liquid, and (c) a burned heater segment made of Al wire due to the high current.

In order to solve the problem of the wire breakage at the silicon die edge, a 0.6μm-deep step was etched around the glass die edge to prevent a direct contact between the metal lines on the glass and silicon die edge. Indeed, as shown in Figure 4.11-a, all the metal lines that cross the Si die edge are continuous after the die-by-die anodic bonding. In addition, instead of aluminium, platinum was selected as the material to form the capacitor top electrodes, metal lines and metal pads on the bonding surface of the glass die with a Ti/W thin film serving as the adhesion layer. The stability of Pt is significantly superior to that of Al. Even after long exposure of Pt to water, the integrity of the electrode was maintained, as evident in Figure 4.11-b. Finally, a thicker aluminium film, 0.75μm instead of 0.15μm, was utilized for the heater fabrication on the silicon wafer. However, the resulting Al line was too thick to allow good anodic bonding. Therefore, the regions on the glass die overlapping the Al lines, formed on the silicon die, were etched to form complimentary grooves about 0.6μm in depth. In this approach, although much thicker, the metal lines created a gap of only 0.15μm between the bonding surfaces of the glass and silicon dies, as in the original fabrication process.

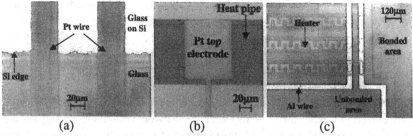

Figure 4.11: Pictures of (a) continuous Pt wires crossing the Si die edge after bonding, (b) a Pt top electrode in contact with the working liquid, and (c) un-burned heater elements in operation under high current.

Similar method was adopted in the formation of metal lines on the glass wafer utilizing the liftoff technique. The mould for the metal lines was transferred to the photoresist film and, prior to metal deposition, the glass wafer was dry etched to form grooves about 0.4μm in depth. Then, a 0.5μm metal layer was deposited followed by the lift off process. This resulted in 0.5μm-thick metal lines with a step of only about 0.1μm protruding from the glass wafer surface. Thus, anodic bonding was accomplished successfully with much thicker metal lines. Consequently, the metal lines were able to sustain the high current required for adequate operation of the heater as evident in Figure 4.11-c. Furthermore, no water leakage from the heat pipes was observed since no gaps existed between the bonded surfaces around the metal lines.

Chapter 5

Thermometry Techniques for Microscale Heat Convection Measurements

Quantitative analysis of heat convection requires measurements of the relevant physical properties. Some integral properties, such as flow rate, can be measured by external means. However, local properties, such as temperature distribution, can be measured with adequate resolution and accuracy only by utilizing integrated sensors. A variety of microsensors have been developed for the study of heat transfer. In this chapter, two examples of microsensors will be described: thermoresistors for temperature distribution and capacitive sensors for void fraction measurements.

5.1   Temperature measurements

The working principle of thermoresistors is based on the temperature dependence of the electrical resistivity of the material. Although the resistivity of both metals and semiconductors vary with temperature, their behaviour is quite different. In metals, the resistivity change is essentially caused by changes in carrier mobility, which typically decreases with temperature because of enhanced phonon scattering. Hence, the resistivity increases almost linearly with increasing temperature over a wide range. In single crystal semiconductors, on the other hand, the resistivity change is mainly due to changes in free carrier concentration that increases exponentially with temperature. Thus, the resistivity decreases exponentially with increasing temperature. While the use of undoped semiconductors for thermoresistors is attractive because of the high temperature sensitivity, sensor parameter control and device stability usually suffer. Therefore, it is more practical to use doped semiconductors, which become quasi-metallic at a sufficiently high doping level [47].

5.1.1   Resistivity of polycrystalline silicon (Poly-Si)

Conduction in polycrystalline silicon films has been extensively studied since they became technologically important more than 20 years ago. It has been recognized that a proper model for the analysis of polycrystalline materials should take into account the existence of multiple crystallite grains in the films [116]. The resistivity of polysilicon is modelled to arise from a series combination of two terms: "bulk" resistivity within a grain, which is controlled by regular phonon and impurity scattering processes, and "barrier" resistivity across the grain boundaries, which is controlled by a thermionic emission process. The resistivity is highly sensitive to light doping concentration in the range between 10^{16}-10^{18}cm^{-3}, which makes it difficult to accurately control film resistivity. Therefore, in this work, moderate doping concentrations between 10^{18} and mid-10^{19}cm^{-3} have been selected for more stable parameter control. For high doping, above 10^{20}cm^{-3}, the resistivity becomes relatively insensitive to doping concentration [81].

A hyperbolic sine dependence of the current, I, on the applied voltage, V, can be derived for a conduction mechanism controlled by thermionic emission to yield the resistance, R, as follows [102]:

$$I = I_0 \sinh \frac{V}{V_0} \cong I_0 \frac{V}{V_0} \quad \Rightarrow \quad R = \frac{V}{I} \cong \frac{V_0}{I_0} \tag{5.1}$$

The approximations apply when the voltage drop across the grain boundaries is small compared with the term kT/q. The parameters I_0 and V_0 are given by:

$$I_0 = 2AA^*T^2 \exp\left(-\frac{qV_{B0}}{kT}\right) \; ; \; V_0 = \frac{2kTN}{q} \tag{5.2}$$

where T is the temperature of the resistor, A is the resistor cross-sectional area perpendicular to the current flow, A^* is the Richardson constant, V_{B0} is the barrier height at each grain boundary, N is the number of grains, q is the electron charge and k is the Boltzmann constant. Once the resistance is known, the resistivity χ is readily available:

$$\chi = \frac{R \cdot A}{L} \tag{5.3}$$

where L is the length of the resistor. Assuming that the conduction in moderately doped polycrystalline resistor is still dominated by thermionic emission, the temperature coefficient of resistance (TCR) is expressed as:

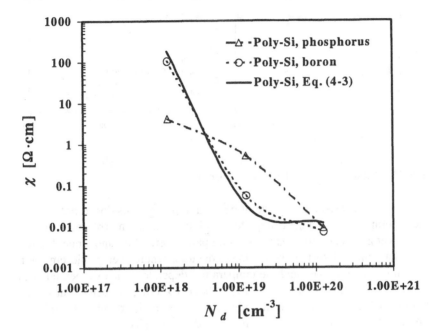

Figure 5.1: Resistivity dependence on doping concentration.

$$\alpha = \frac{d \ln R}{dT} = -\frac{1}{T} - \frac{qV_{BO}}{kT^2} \tag{5.4}$$

The TCR α is negative and nonlinear in T. For moderately doped polycrystalline materials, the grains are only partially depleted and V_{BO} can be approximated by:

$$V_{BO} \cong \frac{Q_T^2}{8q\varepsilon N_d} \tag{5.5}$$

where Q_T is the trap charge per unit grain boundary area, ε is the dielectric constant, and N_d is the doping concentration. Clearly, both V_{BO} and $|\alpha|$ decrease with doping concentration. A comparison between polysilicon resistivity measurements and calculations is shown in Figure 5.1 as a function of doping concentration. For moderate doping level, 10^{18}-10^{19}cm^{-3}, the resistivity of the polysilicon decreases exponentially.

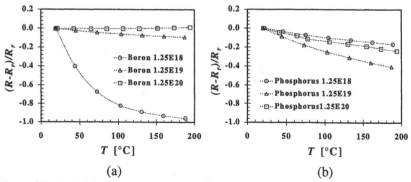

Figure 5.2: Resistivity dependence on temperature.

The relative resistance change of boron and phosphorus-doped polysilicon due to temperature change is depicted in Figures 5.2(a) and (b), respectively, with doping concentration as the varying parameter. At the moderate doping, the behaviour is like undoped semiconductors, where thermionic emission mechanism is dominant and the temperature dependence is exponential with negative TCR. At higher doping level, above 10^{20}cm^{-3}, the behaviour is like metals, where the carrier mobility mechanism is dominant and the temperature dependence is nearly linear with positive TCR.

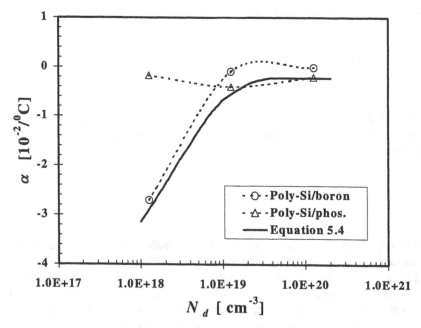

Figure 5.3: TCR dependence on doping concentration.

In general, the resistivity dependence on temperature increases with decreasing doping concentration. The dependence of the TCR on doping concentration is shown in Figure 5.3. For boron-doped film, α approaches zero or a slightly positive value as the concentration increases from 10^{18} to 10^{19}cm^{-3}. The calculated TCR for poly-Si film (Equation 5.4) is in close agreement with the experimental data, and its dependence on the doping concentration is highly non-linear. The largest negative TCR value of boron-doped poly-Si film with concentration of 1.25×10^{18}cm^{-3} is about -0.03/°C.

5.1.2 Temperature sensor design and fabrication

The performance requirements of a sensor dictate its design. Two of the most important requirements are sensitivity and stability. It is well know that lightly-doped films, while less stable, are much more sensitive than highly-doped films that are more stable. In order to satisfy these two conflicting requirements, bridge-like thermoresistors can be constructed on moderately doped films. Shown in Figure 5.4, the two ends of a bridge, serving as signal leads, are heavily doped to give low temperature sensitivity and low electrical resistivity. The "sensing" region at the centre is moderately doped for optimal parameter control, temperature sensitivity and operational stability. It can also be made very small for high spatial resolution.

The fabrication starts with the formation of an electrically insulating layer, followed by the deposition of a 0.5μm-thick undoped polysilicon film. The film is then patterned into dumbbell-shaped resistor bridges. The entire film is moderately implanted with boron at 60KeV, or phosphorus at 160KeV, before the 'lead' portions are heavily doped. The dosage level is about 10^{14}/cm^2 for the moderate and about 5×10^{15}/cm^2 for the heavy doping. The doped film is then annealed at 900°C for 30min. A layer of low-temperature oxide (LTO), 0.5-1.0μm in thickness, is deposited for sensor passivation and contact windows definition. Finally, Al-Si alloy is sputter-deposited, patterned, and sintered at 450°C for 30min in forming gas to provide low-resistance electrical contacts.

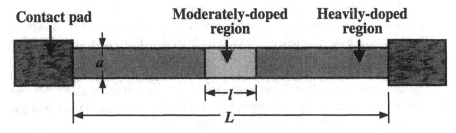

Figure 5.4: A schematic diagram of a dumbbell-shaped thermo-resistor.

5.1.3   Sensor bridge analysis

When current passes through a sensor, electrical power is dissipated and converted into heat by Joule's heating within the sensor. A fraction of this power is stored as internal energy, thus increasing the sensor temperature, and the rest is transferred to the surrounding via conduction, convection and radiation. The energy balance at any point along the sensor bridge can be described by the following equation:

$$\frac{1}{\alpha_b}\frac{\partial T}{\partial t} = \frac{\partial^2 T}{\partial x^2} - \kappa\left(T - \frac{\psi}{\kappa}\right) \tag{5.6}$$

with

$$\kappa = \beta + \gamma - \delta\alpha \quad ; \quad \psi = \beta T_a + \gamma T_w + \delta(1 - \alpha T_0) \tag{5.7}$$

The various coefficients are given by:

$$\alpha_b = \frac{k_b}{c_b \rho_b} \; ; \; \beta = \left(\frac{h}{k_b} + \frac{4\sigma_b T_a^3}{k_b}\right)\left(\frac{2}{a} + \frac{1}{b}\right) \; ; \; \gamma = \frac{k_s}{k_b}\frac{1}{bd} \; ; \; \delta = \frac{J^2 \chi_0}{k_b} \tag{5.8}$$

where a and b are the width and height of the bridge, J is the current density, χ_0 is the resisitivity at a reference temperature T_0; ρ_b, k_b and c_b are the mass density, thermal conductivity and specific heat capacity of the bridge material, respectively; k_s is the thermal conductivity of the insulating layer and d its thickness; T_a and T_w are the ambient and the substrate temperature, respectively; h is the heat-transfer coefficient related to the Nusselt number as follows: $Nu = h \cdot D_h / k_g$, D_h being the bridge hydraulic diameter and k_g the thermal conductivity of the ambient gas. The derivation of the governing equation is detailed elsewhere [127], and it can be solved by separation of variables with proper boundary and initial conditions.

The steady-state temperature distribution along the bridge can be calculated by dropping the time-dependent term in Equation 5.6. However, the equation has to be solved twice for the moderately-doped sensor (I) and the heavily-doped lead region (II), since the resistivity in the two regions is about an order of magnitude apart. In order to simplify the solution, we can select: $T_a = T_w = T_0$. This case corresponds to a situation where both the substrate and the bridge reference temperature are equal to the ambient gas temperature. The governing equation reduces to:

$$\frac{\partial^2 T}{\partial x^2} - \kappa(T - T_0) + \delta = 0 \tag{5.9}$$

The corresponding boundary conditions are:

$$\frac{\partial T_I}{\partial x} = 0 \quad @ \quad x = 0 \tag{5.10-a}$$

$$T_I = T_{II} \quad @ \quad x = l/2 \tag{5.10-b}$$

$$\frac{\partial T_I}{\partial x} = \frac{\partial T_{II}}{\partial x} \quad @ \quad x = l/2 \tag{5.10-c}$$

$$T_{II} = T_0 \quad @ \quad x = L/2 \tag{5.10-d}$$

where l is the length of the centre, moderately doped region, and L is the length of the entire bridge including both the sensor and lead regions.

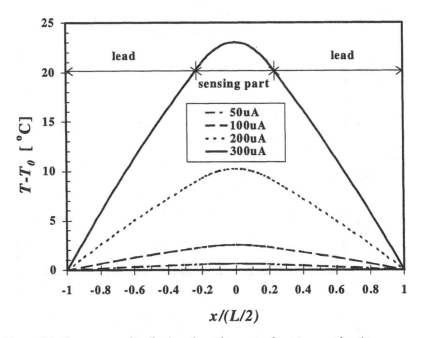

Figure 5.5: Temperature distribution along the sensor due to current heating.

The solution for the two regions is given by:

$$T_I(x) = -A_I l_{cI}^2 + C_1 \left[\exp(x_I / l_{cI}) + \exp(-x_I / l_{cI}) \right]$$ (5.11-a)

and

$$T_{II}(x) = -A_{II} l_{cII}^2 + C_2 \exp(x_{II} / l_{cII}) + C_3 \exp(-x_{II} / l_{cII})$$ (5.11-b)

with

$$A_i = \frac{J^2 \chi_i}{k_b} \quad ; \quad l_{ci} = \left[\frac{2(a+b)h}{abk_b} - \frac{J^2 \chi_i \alpha_i}{k_b} \right]^{-1/2}$$ (5.12)

The subscript $i=I$ stands for the moderately doped and $i=II$ for the heavily doped region. The coefficients C_1, C_2 and C_3 can be obtained by imposing the boundary conditions. The material properties such as the specific heat capacity, thermal conductivity and mass density are assumed to be the same for both regions. The calculated temperature distributions due to different current levels for a poly-Si bridge are plotted in Figure 5.5. As expected, the temperature increases everywhere along the bridge as the current increases. Also, the temperature at the moderately doped centre region of the bridge is the highest since most of the heat dissipation takes place there.

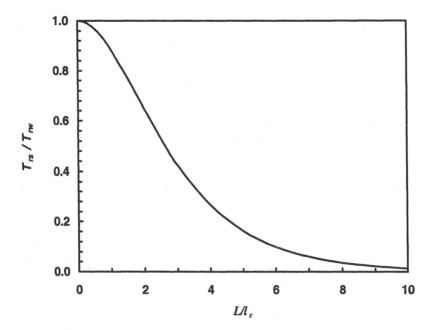

Figure 5.6: Sensor temperature dependence on sensor properties.

The steady-state temperature distribution along the bridge, Equation 5.11, gives rise to a characteristic length scale, l_c. This length scale depends not only on the bridge dimensions, but also on the material properties and on the operating conditions (Equation 5.12). Thus, the bridge temperature strongly depends on l_c as demonstrated in Figure 5.6, where $T_{rs}=(T_s-T_a)/(T_s-T_a)|_{L=\infty}$ and $T_{rw}=(T_w-T_a)/(T_s-T_a)|_{L=\infty}$. The bridge maximum temperature T_s approaches the wall temperature T_w when either the bridge length, L, approaches zero or the characteristic length scale, l_o approaches infinity. In order to enhance the bridge sensitivity, i.e. larger separation between the bridge and the wall temperature, either L should be very large or l_c should be very small.

The unsteady performance of the bridge is investigated by solving Equation 5.6 with the proper boundary and initial conditions. Again, for simplicity we it is assumed that $T_w=T_g=T_0$, radiation is neglected, and the boundary conditions in Equation 5.10 are utilized for all times. The initial condition is uniform temperature everywhere before a step current is applied to the bridge, namely:

$$T_I(x)=T_{II}(x)=T_0 \quad @ \quad t=0 \qquad (5.13)$$

For a uniformly doped bridge with a fixed temperature at both ends, the solution is a multi-mode response, and the eigenvalue of the first mode determines the time constant:

$$\tau_1 = \frac{\tau_{d1}\tau_h}{\tau_{d1}+\tau_h} \qquad (5.14\text{-a})$$

and

$$\tau_{d1} = \frac{\rho_b c_b}{k_b}\left(\frac{l^2}{\pi^2}\right) \quad ; \quad \tau_h = \frac{\rho_b c_b}{\alpha J^2 \chi_0 - 2h(a+b)/ab} \qquad (5.14\text{-b})$$

It can be seen from the above expressions that the time constant is mainly due to two components: a conduction time constant, τ_{d1} which has an l^2 dependence and therefore is small and dominates for short bridges, and a convection time constant, τ_h, which dominates for long bridges. In this case, the conduction time constant is expected to dominate, i.e. $\tau_{d1}<\tau_h$, and thus:

$$\tau_1 = \frac{\tau_{d1}\tau_h}{\tau_{d1}+\tau_h} \approx \frac{\tau_{d1}\tau_h}{\tau_h} = \tau_{d1} \qquad (5.15)$$

It should be noted that this is not a first-order system due to the conduction mechanism, which introduces a higher-order term. Therefore, more than one time constant may exist. For example, the second time constant associated with the second mode is:

$$\tau_2 = \frac{\tau_{d2}\tau_h}{\tau_{d2} + \tau_h} \qquad (5.16\text{-a})$$

where,

$$\tau_{d2} = \frac{\rho_b c_b}{k_b}\left(\frac{l^2}{4\pi^2}\right) \quad ; \quad \tau_h = \frac{\rho_b c_b}{\alpha J^2 \chi_0 - 2h(a+b)/ab} \qquad (5.16\text{-b})$$

Notice that the convection time constant is unchanged, since the terms involved enter the governing equation as first-order system. However, the conduction time constant has changed as this term is of higher order. In the present bridge design, the combined contribution of the moderately- and heavily-doped regions to the time constant is not trivial. Jiang et al. [79] suggested using the same expressions with an effective length of the bridge, l_e, which should be longer than l and smaller than L.

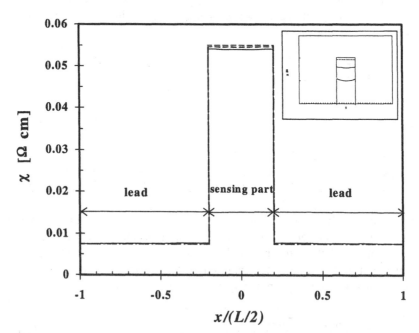

Figure 5.7: Resistivity distribution along the sensor due to current heating.

5.1.4   Temperature microsensor performance

It is difficult to directly measure the temperature distribution along the bridge. However, utilizing the resistivity-temperature data, the bridge resistivity distribution due to the temperature distribution can be calculated. The results are shown in Figure 5.7. The resistivity at the centre is about an order of magnitude higher than that at the two heavily doped leads due to the different doping levels. The resistance of the entire bridge can be calculated for any given current by integrating the resistivity as follows:

$$R = \int_{-L/2}^{+L/2} \frac{\chi(x)}{ab} dx \tag{5.17}$$

The voltage drop across the bridge for any given current level, I, is then given by $V = I \cdot R$. These calculated values are compared in Figure 5.8 with direct I-V curve measurements obtained by a curve tracer. The agreement between the calculations and the measurements is satisfactory. The deviation at high current is probably due to the exclusion of higher-order TCR terms when formulating the Joule's heating term in Equation 5.6. Nevertheless, the first-order TCR model yields reliable results for current levels below 1mA.

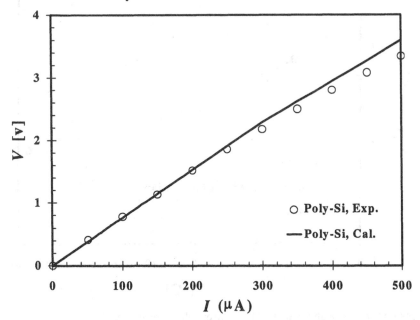

Figure 5.8: Measurement and calculation of a sensor I-V characteristics.

An important aspect of a sensor performance is its stability (drift) during operation and its repeatability over an extended period of time. Resistors based on heavily-doped polycrystalline materials should demonstrate a high level of long term stability since surface effect plays only a minor role in device characteristics, especially with a passivation layer. Obermeier (1986) reported that, at a temperature of 125°C, a drift of less than 0.005 was obtained over a time period of 1000 hours. However, this was for uniformly doped polysilicon resistors. In the present bridge design, the stability may not exhibit similar performance since sharp boundaries between the moderately- and highly-doped regions exist and, as a result, the sensor performance may change over time. Indeed, burn-in has become an important tool in improving the quality of semiconductor devices [222]. Two approaches in burn-in procedure can be effective for the present sensor design: temperature cycling (TC) and duty cycling (DC). In TC approach, the sensor ambient temperature varies in an appropriate range, and a time period of about 30min is required for each temperature. This process should be repeated 5-10 times. In DC approach, an appropriate current is applied to the sensor for a long time. A time period of more than 100hr is typically required. Naturally, if both approaches are combined simultaneously, the equivalent hours can be compressed into a shorter period of time.

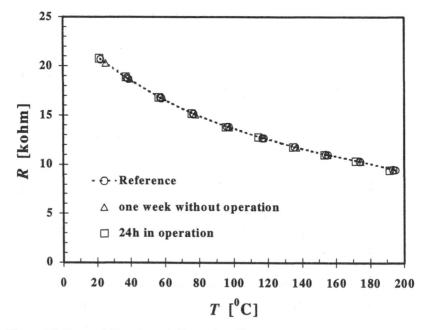

Figure 5.9: Repeatability of a typical sensor calibration curve.

The calibration of a polysilicon sensor 4μm×4μm×0.5μm in dimensions, after a burn-in procedure of 5 cycles TC at a temperature range of 20-200°C and 50hr DC with 50μA, is summarized in Figure 5.9. The sensor resistance was first measured for reference as a function of the ambient oven temperature, where the sensor had been placed. The calibration procedure was then repeated twice: after one week during which time no current was passed through the sensor, and again after 24hr of maintaining a constant current of 50μA through the sensor. All the calibration curves are nonlinear with a negative TCR, typical to phosphorus doping. Moreover, the difference among the three calibration curves is less than 1%. Hence, drift of less than 0.01 in the resistance of the bridge sensors, with or without operation, can be achieved after appropriate burn-in process.

The time constant is experimentally determined by passing a step current through the bridge in still air and observing the voltage response on an oscilloscope. Typical voltage-waveform responses of a bridge with negative TCR to a step current are shown in Figure 5.10. The initial rise time in the heating process, Figure 5.10(a), is due to the electrical time constant as it takes a finite time for the current to increase from zero to the desired value. Initially, the bridge is at room temperature and its resistance is high; thus, the voltage output attains its maximum value. However, as the electrical power is dissipated into heat, the thermal time constant comes into play. Since the bridge resistance decreases while the temperature is increasing, as the current is held constant, the voltage output drops from its peak to a steady-state value. A similar behaviour is observed in the cooling process when the current is turned off as shown in Figure 5.10(b). However, if the voltage output drops to zero, i.e. no bias voltage, the voltage overshoot is not as prominent as in the heating process.

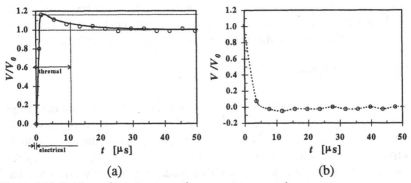

(a)   (b)

Figure 5.10: Bridge voltage response due to a step current input.

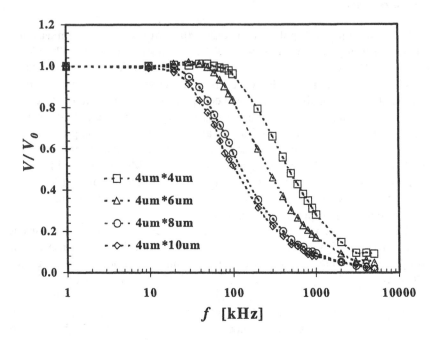

Figure 5.11: Sensor length effect on the frequency response.

Although the time constant and the frequency response are closely related, the frequency response in many instances is of more interest. Both depend on material properties (ρ_b, c_b, k_b, λ, χ_{0}), sensor geometry (a, b, l) and operating conditions (J, $h(Re)$). The frequency response is depicted in Figure 5.11 for bridge structures 4μm wide with varying sensor length. It is expected that the effective length of the bridge will increase as the length of the center, moderately-doped region increases. The dominant conduction time constant will then increase as well leading to a slower response. Indeed, the corner frequency for poly-Si bridges increases, indicating a smaller time constant, as the length of the moderately doped-region decreases.

5.2   Void-fraction measurements

Physical systems involved in convective heat transfer may include single-phase fluid flow, with liquid or gas only. Other systems, however, could be designed to utilize phase change of the working fluid either from liquid to vapour (evaporation) or from vapour to liquid (condensation). The systems then will feature two-phase flow, either partially or entirely. In such a case, the ratio between the vapour and liquid fraction, defined as the void fraction (volume ratio) or quality (mass ratio), becomes of primary importance. A complete characterization of a thermal system with two-phase flow convective heat transfer would then require information about the instantaneous void fraction or quality distribution. Simple capacitance or impedance sensors have been used since the 1960s for measurement of two-phase fluid fractions. The advantage of using capacitance transducers is their high speed, low cost, and simple principle-of-operation.

5.2.1   Capacitance sensor for void fraction measurement

In single-phase flow the capacitance between two parallel plate electrodes is simply given by:

$$C = \frac{\varepsilon \varepsilon_0 A}{h} \tag{5.18}$$

where ε_0 is the vacuum permittivity, while A, h and ε are the overlap area, the separation distance, and the dielectric constant of the medium between the capacitor two electrodes. Indeed, capacitive microsensors that are sensitive to displacement, i.e. changes in A or h, are widely used although usually not to measure displacement but other related measurands. For example, displacement is measured capacitively in many micromechanical structures such as cantilevers, membranes, and resonant flexures, to measure indirectly acceleration, force/torque, or pressure/stress. In these applications, the media between the sensor electrodes stays the same, i.e. $\varepsilon\varepsilon_0 = const.$ However, taking advantage of the great disparity between the dielectric constants of liquids and gases, capacitance sensors can also be used for void-fraction measurement being sensitive to changes in ε with constant A and h.

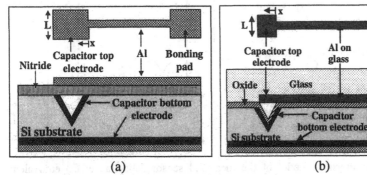

Figure 5.12: Schematic cross-sections for fabricating a capacitance sensor using: (a) CMOS-compatible and (b) glass-based technology.

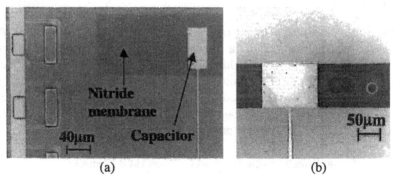

(a)   (b)

Figure 5.13: Close-up pictures of a fabricated capacitance sensor top electrode using: (a) CMOS-compatible and (b) glass-based technology.

Two different technologies that have been utilized to integrate capacitance sensors in thermal microsystems, i.e. CMOS-compatible and glass-based, are respectively illustrated in Figures 5.12-a and b. Close-up pictures of the fabricated sensors are correspondingly shown in Figure 5.13. In both techniques, the capacitor bottom electrode is created by heavily doping the microchannel bottom surface and the wafer backside. A thin metal layer is sputtered and patterned on the channel cover, either the nitride membrane (Figure 5.13-a) or the glass wafer (Figure 5.13-b), to form a rectangular top electrode. Both the nitride film and the glass substrate provide good insulation between the top electrodes of the distributed capacitors. The detailed fabrication process is discussed in Chapter 4.

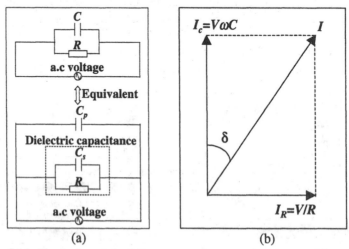

(a)   (b)

Figure 5.14: Physical model of the measured sensor capacitance: (a) equivalent circuit including the resistance, sensor and parasitic capacitance, and (b) corresponding phase diagram showing the dielectric capacitance with tangent loss.

The capacitance microsensor is regarded as a dielectric capacitor when the space between the electrodes is filled with water. There is no perfect dielectric material and loss tangent, $\tan\delta=1/(\omega RC)$, must be recognized [21]. This loss value is a direct expression of the existence of a resistance, R, associated with the dielectric capacitance, C. Therefore, the dielectric capacitor is modelled using the parallel model, in which C_s is in parallel with R [9]. In addition, a parasitic capacitance C_p, due to an overlap between the metal interconnects and the silicon substrate, is also present in parallel with the dielectric capacitance as shown in Figure 5.14-a. Thus, the equivalent model includes the measured capacitance C in parallel to R, and the relationship between C and R is shown in Figure 5.14-b.

5.2.2   Parameters affecting the capacitance sensor measurements

Equation 5.18 is the basis for using capacitance as a sensing principle. Capacitors can be used to measure any property whose combined effect results in variations of either A, h, or ε. The sensor capacitance is directly proportional to the overlap area A between the sensor electrodes with a constant separation distance h as long as the dielectric constant of the medium ε does not change. Experiments have been conducted in a microchannel integrated with capacitors having the same width but varying in length. Indeed, the measured capacitance, shown in Figure 5.15, increases linearly with the sensor length for either water or air as the medium between the sensor electrodes.

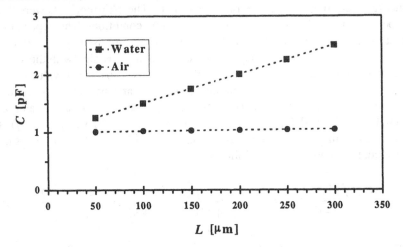

Figure 5.15: Capacitance dependence on the sensor length, with a constant sensor width, for pure water and pure air between the capacitor electrodes.

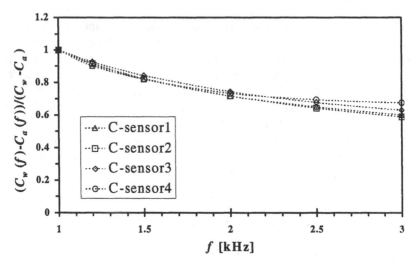

Figure 5.16: Normalized sensor capacitance dependence on the measurement frequency for D.I. water at 20°C.

The sensor capacitance under constant temperature is a function of the medium dielectric constant, which depends on the measurement frequency. Thus, the capacity dependence on the measurement frequency f for various sensors is demonstrated in Figure 5.16. The capacitance is first measured in 100% air at room temperature of $T_0=20°C$ using the LCR meter. The capacitance in air, C_a, was found to be weakly dependent of the measurement frequency in the range $f=1-3kHz$. The measurements were then repeated for 100% water, $C_w(f)$, under the same conditions. The capacitance measurements at $f=1kHz$ for air, C_a, and water, C_w, are used as the reference values. The parasitic capacitance is modelled to be in parallel with the sensor capacitance (Figure 5.14-a). Hence, each measured capacitance value can be viewed as the sum of two components: sensing and parasitic capacitance. Correspondingly, the reference sensor capacitances in air and water are denoted by C_{sa} and C_{sw}, while the sensor capacitance for water is $C_{sw}(f)$. In order to eliminate the effect of parasitic capacitance, C_p, the measured capacitance is normalized as follows:

$$\frac{C_w(f)-C_a(f)}{C_w-C_a} = \frac{\left[C_{sw}(f)+C_p\right]-\left[C_{sa}(f)+C_p\right]}{\left(C_{sw}+C_p\right)-\left(C_{sa}+C_p\right)} \cong \frac{C_{sw}(f)}{C_{sw}} \quad (5.19)$$

The experimental results shown in Figure 5.16 indicate that the capacitance value for 100% water decreases with increasing frequency from 1kHz to 3kHz. This means that the dielectric constant of D.I. water also decreases in this frequency range, although the dielectric constant of water is expected to be independent of frequency up to 100MHz [61].

The sensor capacitance is directly proportional to the dielectric constant of the working liquid, which in-turn depends on the temperature. The capacity measurements for 100% water at a constant frequency of 1kHz, $C_w(T)$, are summarized in Figure 5.17 for a temperature range $T-T_0=0-50°C$. The corresponding capacitance for 100% air, C_a, was found to be independent of temperature at the same range. Again, to eliminate the parasitic capacitance effect, the measured capacitance is normalized as:

$$\frac{C_w(T)-C_a}{C_w-C_a}=\frac{\left[C_{sw}(T)+C_p\right]-\left(C_{sa}+C_p\right)}{\left(C_{sw}+C_p\right)-\left(C_{sa}+C_p\right)}\cong\frac{C_{sw}(T)}{C_{sw}} \tag{5.20}$$

The measured capacitance of all sensors in Figure 5.17 increases monotonically with increasing temperature. It is known that the dielectric constant of water is inversely proportional to the absolute temperature, $\varepsilon_w(T)\propto1/T$ [45]. For example, the dielectric constant of water at 70°C is 85% of its value at 20°C. Hence, the dielectric constant of water and, consequently, the sensor capacitance should decrease with increasing temperature. However, repeated experiments for a variety of sensors in different devices always resulted in the same trend of increasing sensor capacitance with increased temperature.

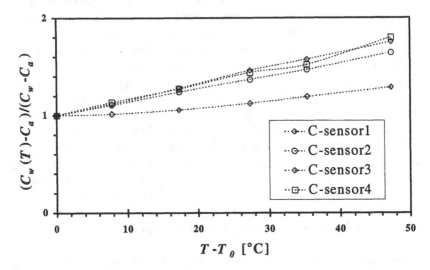

Figure 5.17: Normalized sensor capacitance dependence on temperature for D.I. water at 1kHz measurement frequency ($T_0=20°C$).

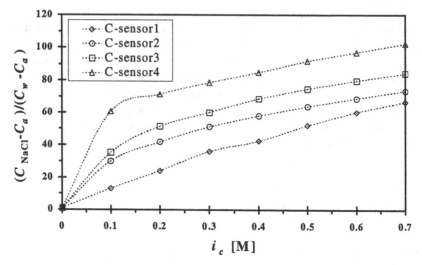

Figure 5.18: Normalized sensor capacitance dependence on the ion concentration of NaCl in D.I. water at 20°C and 1kHz.

Different values were obtained for repeated measurements of sensor capacitance under seemingly identical conditions, with 100% water, and this was attributed to different ion concentration in the working liquid. Since this process was random, a controlled experiment was conducted to verify the dependency of the sensor capacitance on ion concentration. Sodium chloride was added to the DI water, changing its ion concentration i_c from nearly zero to 0.7M. The capacitance of each sensor along the heat pipe was measured at 1kHz and 20°C, and the results are shown in Figure 5.18. Evidently, the capacitance of all sensors increases monotonically with increasing ion concentration for the tested range. Hence, since all other parameters are kept unchanged, the dielectric constant also increases significantly with ion concentration. This is in direct contrast with previous studies reporting that the dielectric constant of aqueous sodium chloride is constant for concentration of sodium chloride ranging from 0.003M to 0.7M [19].

The microsensor capacitance for 100% water is found to depend on the measurement frequency, temperature and ion concentration. The effects of these three parameters are different from expected trends reported in the literature, and further research is required to study these phenomena.

5.2.3 Calibration of an integrated capacitance sensor

The dielectric constant of water is very different from that of air or vapour. Hence, the measured capacitance is sensitive to the air/water volume ratio of the fluid mixture between the sensor electrodes. Since the dielectric constant of vapour is almost equal to that of air, upon proper calibration, the capacitive sensor can be used for void-fraction measurements of the two-phase flows inside a microduct during operation. The calibration of the capacitance sensor is very tricky. The two extreme points of single-phase air or water are clear and easy to measure. This would have been sufficient to establish the entire calibration curve if the capacitance was a linear function of the air/water volume ratio. This is not the case, and the capacitance could be different even for the same air/water ratio with different phase distribution in space. Therefore, more calibration points are required, for which it is difficult to control the air/water volume ratio of the fluid between the electrodes in a microduct due to the strong capillary forces.

A possible calibration technique applied in triangular microchannels is described [109]. Before calibration, the capacitance value in air medium C_a, at 1kHz and 20°C, is registered. The inlet of the microchannel is then connected to a syringe, and the syringe presses the D.I. water into the microchannel to fill it up. The water inside the channel gradually evaporates through the open channel outlet forming a water-air interface, which propagates slowly along the microchannel. Before the interface arrives at a given sensor, as shown in Figure 5.19-a, the capacitance value for 100% D.I. water C_w is measured. As the tip of the air-water interface passes through the capacitance senor, the exact position of the interface cannot be observed since the sensor electrode is opaque. Still, the capacitance value for various air/water mixtures between the sensor electrodes has to be recorded to obtain a calibration curve. Only after the interface tip has passed through the sensor, such that it can be observed beyond the other edge of the electrode as shown in Figure 5.19-b, can the void fraction be estimated with its corresponding measured capacitance.

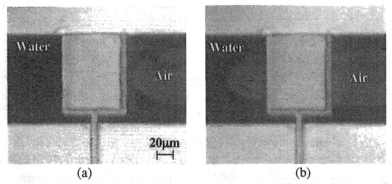

(a)   (b)

Figure 5.19: Close up pictures showing the water-air interface: (a) before, and (b) after the interface has passed through the capacitor top electrode.

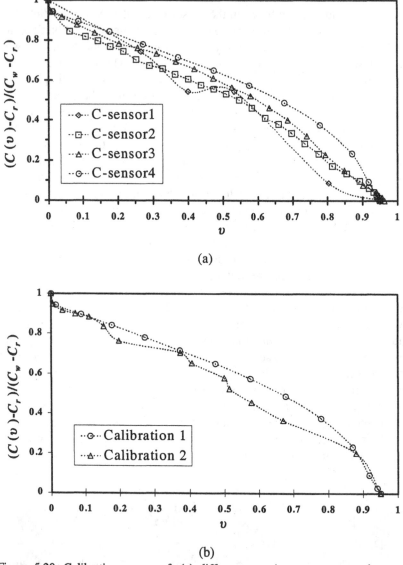

(a)

(b)

Figure 5.20: Calibration curves of: (a) different capacitance sensors at the same time, and (b) same capacitor at different times.

As the air-water interface travels through the duct from one end to the other, liquid meniscuses remain trapped in the corners of the triangular channel. Under this condition, the volume fraction of the air underneath the sensor is estimated to be about 95%, and the corresponding measured capacitance is used as the reference value C_r. The measured capacitance C, normalized by the reference capacitance C_r, is plotted in Figure 5.20 as a function of the estimated void fraction v. The normalized calibration curves for sensors with different area collapse together, as shown in Figure 5.20-a, exhibiting an almost linear decrease of the capacitance with increasing air volume fraction from 0 to 95%. Since the calibration environment is not perfectly clean, contaminating ions from the surroundings affect the ion concentration in the water. Therefore, the ion concentration within the working liquid is different and cannot be controlled during each calibration process, resulting in different capacitance measurements for the same void fraction estimates during repeated calibrations. However, as shown in Figure 5.20-b, normalizing the measured capacitance with the reference, C_r, rather then the air value, C_a, provides similar curves at different calibration times. The resulting calibration curves are then independent of the ion concentration in the working liquid. It is interesting to note that as the meniscuses dry out, the capacitance values of all sensors suddenly drop to the corresponding air values, C_a, such that no additional data point can be obtained for air volume fraction in the range 95-100%.

Chapter 6

Steady, Single-Phase Heat Convection in Micro Ducts

Flows completely bounded by solid surfaces are called internal flows, and they include flows through ducts, pipes, nozzles, diffusers, etc. External flows are flows over bodies in an unbounded fluid. Flows over a plate, a cylinder or a sphere are examples of external flows, and they are not within the scope of this chapter. Only internal flows, in either liquid or gas phase, in micro ducts will be discussed emphasizing size effects, which may potentially lead to a different behaviour in comparison with similar flows in macro ducts. For the practical application of single-phase heat convection in micro-ducts, e.g. heat exchangers, the friction factor and the heat transfer coefficient are the parameters of prime interest. As long as material properties, such as viscosity, are nearly constant over the temperature range of operation, the friction factor can be predicted analytically without invoking the energy equation. The extensive experimental and theoretical research has matured to a point where the size effects are well understood. Unfortunately, this is not the case with respect to the heat transfer coefficient, mainly due to lack of reliable experimental data needed to confirm or refute the numerous theoretical analyses and numerical simulations.

6.1   Flow structure

Viscous flow regimes are classified as laminar or turbulent on the basis of flow structure. In the laminar regime, flow structure is characterized by smooth motion in laminae, or layers. The flow in the turbulent regime is characterized by random, three-dimensional motions of fluid particles superimposed on the mean motion. These turbulent fluctuations enhance the convective heat transfer dramatically. However, turbulent flow occurs in practice only when the Reynolds number is greater than a critical value, Re_{cr}. The critical Reynolds number depends on the duct inlet conditions, surface roughness, vibrations imposed on the duct walls and the geometry of the duct cross-section. Values of Re_{cr} for various duct cross-section shapes have been tabulated elsewhere [16]. In practical applications though, the critical Reynolds number is estimated to be:

$$Re_{cr} = \frac{\rho U_a D_h}{\mu} \cong 2300 \qquad (6.1)$$

where U_a is the mean flow velocity, and $D_h=4A/S$ is the hydraulic diameter; A and S being the cross-section area and the wetted perimeter, respectively. Microchannels are typically larger than 1000μm in length and with a hydraulic diameter of about 10μm. The mean velocity for gas flow under pressure drop of about 0.5MPa is less than 100m/s, and the corresponding Reynolds number is less than 100. The Reynolds number for liquid flow will be even smaller due to the much higher viscous forces. Thus, in most applications, the flow in microchannels is expected to be laminar. Turbulent flow may develop in short channels with large hydraulic diameter under high-pressure drop and, therefore, is not likely to develop in microchannels.

6.2   Entrance length

When a viscous fluid flows in a duct, a velocity boundary layer develops along the inside surfaces of the duct. The boundary layer fills the entire duct gradually as sketched in Figure 6.1. The region where the velocity profile is developing is called the hydrodynamics entrance region, and its extent is the hydrodynamic entrance length. An estimate of the magnitude of the hydrodynamic entrance length, L_h, in laminar flow in a duct is given by [182]:

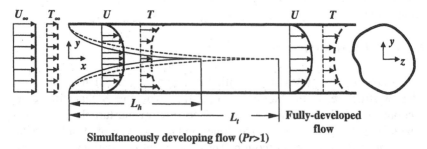

Simultaneously developing flow (*Pr*>1)

Figure 6.1: A schematic of hydrodynamically and thermally developing flow, followed by hydrodynamically and thermally fully-developed flow.

$$\frac{L_h}{D_h} = 0.056Re \tag{6.2}$$

The region beyond the entrance region is referred to as the hydrodynamically fully-developed region. In this region the boundary layer completely fills the duct and the velocity profile becomes invariant with the axial coordinate. If the walls of the duct are heated (or cooled), a thermal boundary layer will also develop along the inner surfaces of the duct, shown in Figure 6.1. At a certain location downstream from the inlet, the flow becomes thermally fully-developed. The thermal entrance length, L_t, is then the duct length required for the developing flow to reach fully-developed condition. The thermal entrance length for laminar flow in ducts varies with the Reynolds number, Prandtl number and the type of the boundary condition imposed on the duct wall. It is approximately given by:

$$\frac{L_h}{D_h} \cong 0.05RePr \tag{6.3}$$

More accurate discussion on thermal entrance length in ducts under various laminar flow conditions can be found elsewhere [182].

In most practical applications of microchannels, the Reynolds number is less than 100 while the Prandtl number is on the order of 1. Thus, both the hydrodynamic and thermal entrance lengths are less than 5 times the hydraulic diameter. Since the length of microchannels is typically two orders of magnitude larger than the hydraulic diameter, both entrance lengths are less than 5% of the microchannel length and can be neglected.

6.3   Governing equations

Representing the flow in rectangular ducts as flow between two-parallel plates, the 2-D governing equations can be simplified as follows [177]:

Continuity,

$$\frac{\partial(\rho u)}{\partial x} + \frac{\partial(\rho v)}{\partial y} = 0 \tag{6.4}$$

x-momentum,

$$\frac{\partial(\rho uu)}{\partial x} + \frac{\partial(\rho vu)}{\partial y} = -\frac{\partial P}{\partial x} + \mu\left(\frac{\partial^2 u}{\partial x^2} + \frac{\partial^2 u}{\partial y^2}\right) + \frac{\mu}{3}\frac{\partial}{\partial x}\left(\frac{\partial u}{\partial x} + \frac{\partial v}{\partial y}\right) \tag{6.5}$$

y-momentum,

$$\frac{\partial(\rho uv)}{\partial x} + \frac{\partial(\rho vv)}{\partial y} = -\frac{\partial P}{\partial y} + \mu\left(\frac{\partial^2 v}{\partial x^2} + \frac{\partial^2 v}{\partial y^2}\right) + \frac{\mu}{3}\frac{\partial}{\partial y}\left(\frac{\partial u}{\partial x} + \frac{\partial v}{\partial y}\right) \quad (6.6)$$

Energy,

$$u\frac{\partial T}{\partial x} + v\frac{\partial T}{\partial y} = \frac{k}{\rho c_p}\left(\frac{\partial^2 T}{\partial x^2} + \frac{\partial^2 T}{\partial y^2}\right) +$$

$$+ \frac{2\mu}{\rho c_p}\left[\left(\frac{\partial u}{\partial x}\right)^2 + \left(\frac{\partial v}{\partial y}\right)^2 + \frac{1}{2}\left(\frac{\partial u}{\partial y} + \frac{\partial v}{\partial x}\right)^2\right] \quad (6.7)$$

6.4   Fully developed gas flow forced convection

An analytical solution of Equations (6.4)-(6.7) is not available. Some solutions can be obtained upon further simplification of the mathematical model. Indeed, incompressible gas flows in macro ducts, with different cross sections, subjected to a variety of boundary conditions are available [182]. However, the important features of gas flow in micro ducts are mainly due to rarefaction and compressibility effects. Two more effects due to acceleration and non-parabolic velocity profile were found to be of second order compared to the compressibility effect [240]. The simplest system for demonstration of the rarefaction and compressibility effects is the two-dimensional flow between parallel plates separated by a distance H, with L being the channel length ($L/H \gg 1$). If $MaKn \ll 1$, all streamwise derivatives can be ignored except the pressure gradient, which is the driving force. The momentum equation for iso-thermal flow reduces to:

$$-\frac{dP}{dx} + \mu\frac{d^2 u}{dy^2} = 0 \quad (6.8)$$

with the symmetry condition at the channel centerline, $y=0$, and the slip boundary conditions at the walls, $y=\pm H/2$, as follows:

$$\frac{du}{dy} = 0 \qquad @ \qquad y = 0 \qquad\qquad (6.9\text{-a})$$

$$u = -\lambda\frac{du}{dy}\bigg|_{y=H/2} \qquad @ \qquad y = \pm H/2 \qquad\qquad (6.9\text{-b})$$

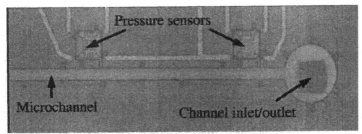

Figure 6.2: A picture of a microchannel, about 1μm high and 40μm wide, integrated with pressure microsensors.

Integration of Equation (6.8) twice with respect to y, assuming $P=P(x)$, yields the following velocity profile [3]:

$$u(y) = -\frac{H^2}{8\mu}\frac{dP}{dx}\left[1-\left(\frac{y}{H/2}\right)^2 + 4Kn(x)\right] \qquad (6.10)$$

where $Kn(x)=\lambda(x)/H$. The streamwise pressure distribution, $P(x)$, calculated based on the same model is given by:

$$\frac{P(x)}{P_o} = -6Kn_o + $$

$$+ \sqrt{\left(6Kn_o + \frac{P_i}{P_o}\right)^2 - \left[\left(\frac{P_i^2}{P_o^2}-1\right)+12Kn_o\left(\frac{P_i}{P_o}-1\right)\right]\left(\frac{x}{L}\right)} \qquad (6.11)$$

where P_i is the inlet pressure, P_o is the outlet pressure, and Kn_o is the outlet Knudsen number. It is difficult to verify experimentally the cross-stream velocity distribution, $u(y)$, within a microchannel. However, detailed pressure measurements have been reported [113,167]. A picture of a microchannel integrated with pressure sensors fabricated to carry out such experiments is shown in Figure 6.2. Pressure measurements reveal that the calculated, non-linear pressure distributions based on Equation (6.11) are in close agreement with the experimental data as demonstrated in Figure 6.3 [240]. The non-linearity is a direct result of the compressible effect, different from incompressible flow with a constant pressure gradient along the channel (under fully-developed flow condition).

The mass flow rate Q_m as a function of the inlet and outlet conditions is obtained by integrating the velocity profile in Equation (6.10) with respect to x and y as follows:

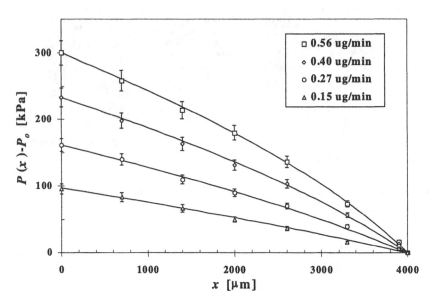

Figure 6.3: A comparison between calculated (solid lines) and measured (symbols) streamwise pressure distributions for various nitrogen flow rates (H=0.53μm).

$$Q_m = \frac{H^3 W P_o^{\,2}}{24 \mu RTL} \left[\left(\frac{P_i}{P_o} \right)^2 - 1 + 12 Kn_o \left(\frac{P_i}{P_o} - 1 \right) \right] \qquad (6.12)$$

where W is the width of the channel, T is the ambient temperature, and R is the specific gas constant. Note the non-linear dependence of the mass flow rate on the pressure drop (or pressure ratio), which is a compressible rather than slip flow effect. The model also suggests that the mass flow rate is inversely proportional to the ambient temperature. Mass flow rate measurements (symbols) in a microchannel about 0.53μm in height using three different working gasses: nitrogen, argon and helium, are compared with calculations based on Equation (6.12) (solid lines) in Figure 6.4. The good agreement suggests that this simple model accounts properly for the compressibility and rarefaction effects, which are not negligible for gas flow in microchannels. Furthermore, the flow rate dependence on the ambient temperature, ranging from 20 to 60°C, has been verified experimentally as demonstrated in Figure 6.5 [80].

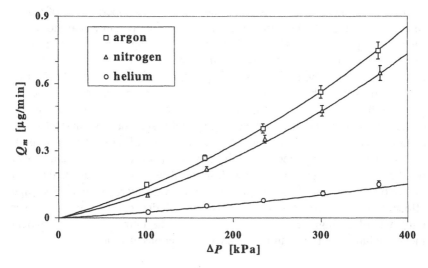

Figure 6.4: A comparison between calculated (solid lines) and measured (symbols) mass flow rate for various working gases (H=0.53μm, W=40μm).

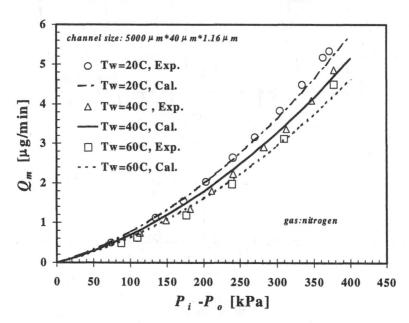

Figure 6.5: A comparison between calculated (solid lines) and measured (symbols) mass flow rate for nitrogen with varying ambient temperature (H=1.16μm, W=40μm).

In some applications, e.g. cooling, the actual wall temperature is of major concern. The overall slip-flow effect on the wall temperature may not clear since the slip flow conditions include two competing effects. The velocity slip at the wall increases the flow rate, thus enhancing the cooling efficiency. On the other hand, the temperature jump at the boundary acts as a barrier to the flow of heat to the gas, thus reducing the cooling efficiency. In order to highlight these slip flow effects, the operation of a heat exchanger consisting of multiple microchannels embedded in a solid substrate is analysed. The conjugate heat transfer problem can be greatly simplified by assuming both hydrodynamically and thermally fully-developed flow. The physical model for the heat exchanger operating with uniform heat flux, q'', is sketched in Figure 6.6. If the system is comprised of numerous microchannel cells, edge effects can be neglected. Consequently, the solution can be computed over only half of a single cell due to symmetry. The corresponding analytical model includes the momentum and energy equations as follows:

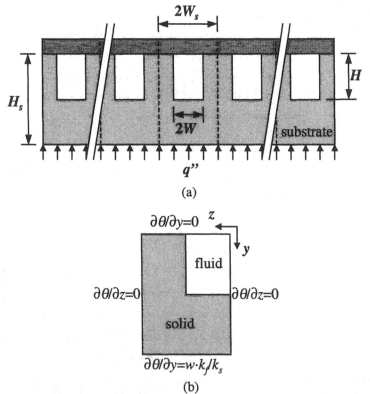

Figure 6.6: Physical model of the micro heat exchanger: (a) schematic cross-section, and (b) the corresponding computational domain.

$$\frac{\partial^2 u}{\partial y^2} + \frac{\partial^2 u}{\partial z^2} = \frac{dP}{dx} \tag{6.13}$$

$$\frac{\partial^2 \theta}{\partial y^2} + \frac{\partial^2 \theta}{\partial z^2} = \begin{cases} u/w & \text{fluid} \\ 0 & \text{solid} \end{cases} \tag{6.14}$$

where $u(y,z)$ is the velocity normalized by the mean velocity, U_a, and $\theta(y,z)$ is the temperature normalized as follows:

$$\theta(y,z) = \frac{k_f}{Wq''}[T(x,y,z) - T_B(x)] \tag{6.15}$$

The fluid-bulk temperarure T_B is given by:

$$T_B(x) = \frac{1}{WHU_a} \int_0^W \int_0^H U(y,z)T(x,y,z)dydz \tag{6.16}$$

In order to highlight the slip flow effect, Equations (6.8)-(6.9) have been solved numerically for silicon as the substrate material and air as the working fluid assuming either continuum or slip flow regime. The simplified boundary conditions are:

$$\partial\theta/\partial z = 0 \quad @ \quad z{=}0,W_s \quad \text{(symmetry)} \tag{6.17-a}$$

$$\partial\theta/\partial y = 0 \quad @ \quad y{=}0 \quad \text{(adiabatic)} \tag{6.17-b}$$

$$\partial\theta/\partial y = 0 \quad @ \quad y{=}H_s \quad \text{(constant heat flux)} \tag{6.17-c}$$

The calculated velocity distributions with ($Kn{>}0$) and without velocity slip at the wall ($Kn{=}0$) are plotted in Figure 6.7. The finite velocity at the wall is evident, resulting in higher flow velocity everywhere. Consequently, due to the size effect, the mass flow rate is higher than the rate predicted by the classical theory ($Kn{=}0$) under the same pressure gradient. The calculated normalized temperature profiles, $\theta(y,z)$, for the continuum and slip-flow regime are plotted in Figure 6.8. Again, the temperature jump at the wall is clear, resulting in higher wall temperature for the slip flow ($Kn{>}1$) compared with predictions based on the classical theory under the same heat flux. This means that the slip flow effect is equivalent to an insulation layer, between the solid and fluid, with vanishing thickness but having a finite temperature difference across the layer.

u (y,z)

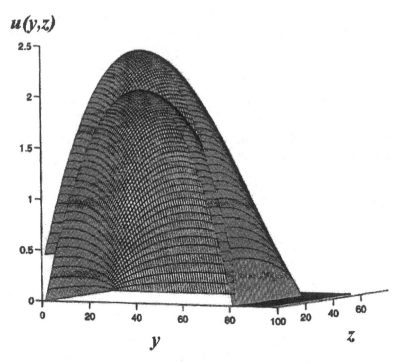

Figure 6.7: Velocity distributions for the hydrodynamically fully-developed flow with and without the velocity-slip boundary condition.

θ (y,z)

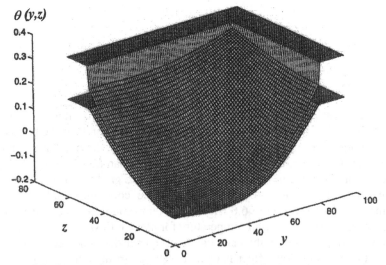

Figure 6.8: Temperature distributions for the thermally fully-developed flow with and without the temperature-jump boundary condition.

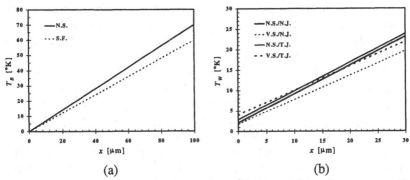

Figure 6.9: Streamwise evolution of the: (a) fluid-bulk, and (b) wall temperature for the hydrodynamically and thermally fully-developed flow.

The fluid-bulk and wall temperatures for continuum and slip flow are plotted in Figure 6.9. In all cases, both temperatures increase linearly with the downstream distance, consistent with the thermally fully-developed assumption. Due to the insulation-like effect of the temperature jump boundary condition, the bulk fluid temperature for the slip flow is lower than the fluid temperature of the continuum flow. This means that less heat per unit mass is being transported by the slip flow, resulting in poorer cooling efficiency. However, the slip flow rate under the same pressure gradient is higher. In order to highlight the two competing effects of the slip flow on the cooling efficiency, the wall temperature is plotted in Figure 6.9-b for four different combinations of boundary conditions: (i) no velocity-slip and no temperature-jump (continuum flow), (ii) no velocity-slip but with temperature-jump, (iii) with velocity-slip but no temperature-jump, and (iv) with velocity-slip and with temperature-jump (slip flow). The temperature jump results in a shift of the wall temperature distribution to a higher level, with the same gradient, regardless of the velocity boundary condition. On the other hand, the velocity slip decreases the wall temperature gradient regardless of the temperature boundary condition. When comparing the slip with the continuum flow, the slip wall temperature is initially higher due to the temperature jump. Further downstream, due to the higher mass flow rate, the wall temperature for slip flow is lower than that for the continuum flow. The net result of these competing effects, i.e. the cross-over steamwise distance, depends on the specific material properties and specific geometry of the system. For simplicity, this exercise has been conducted under the assumption of hydro-dynamically fully-developed flow, which implies incompressible flow. However, the mass flow rate and pressure distribution data indicate that compressibility is the most dominant effect for gas flow in a microchannel. Therefore, the governing equations have to be solved for compressible flow to properly evaluate the slip effect on the heat transfer.

The microchannel flow temperature distribution depends on the heat flux and the corresponding boundary conditions, and limited theoretical work has been conducted [14,55]. However, closed-form analytical solutions in general are still not available. Numerical simulations of Equations (6.4)-(6.7) have been performed to study the effects of gas compressibility and rarefaction in microchannels [90]. The compressible forms of the momentum and energy equations have been solved, with velocity slip and temperature jump boundary conditions (Equation 2.12), in a parallel plate channel for both uniform wall temperature and uniform wall heat flux conditions. The equations were solved simultaneously with the equation of state, since the density is a function of pressure and temperature, using the control volume finite difference scheme. The computed pressure distributions are in good agreement with predictions based on Equation (6.11). This necessarily means that the entrance length, not accounted for in deriving the model Equation (6.8), is negligible as conjectured in Section 6.2. Although the entrance length increases with the Knudsen number, due to the rarefaction effect, flow properties settled within less than 0.1% of the channel length. The friction coefficient decreases with the Knudsen number due the velocity slip at the wall as expected. Once the temperature field is obtained, the heat transfer coefficient or its dimensionless form, Nusselt number, can be computed. For the case of uniform wall temperature, Nu_T is calculated as,

$$Nu_T = \frac{\partial T / \partial y \big|_{wall}}{T_{wall} - T_B} \tag{6.18}$$

and for the case of uniform heat flux, q'', Nu_H is obtained as,

$$Nu_H = \frac{q'' D_h}{k_f \left(T_{wall} - T_B\right)} \tag{6.19}$$

where T is the dimensioness temperature, T_B is the fluid-bulk temperature, D_h is the hydraulic diameter, and k_f is the fluid thermal conductivity. The streamwise evolution of the Nusselt number as a function of the Knudsen number for both conditions, uniform wall temperature and uniform wall heat flux, is summarized in Figure 6.10. Clearly, in both cases, the heat transfer rate from the wall to the gas flow decreases while the entrance length increases due to the rarefaction effect, i.e. increasing Knudsen number. It is worth noting again that the Nusselt number approaches the fully-developed value within a distance on the order of the channel height. In both cases, the Nuseelt number decreases from about 8 for $Kn=0$ to about 5 for $Kn=0.1$. Interestingly, the Nusselt number downstream of the entrance region is independent of the Reynolds number in the range $0.01>Re>10$.

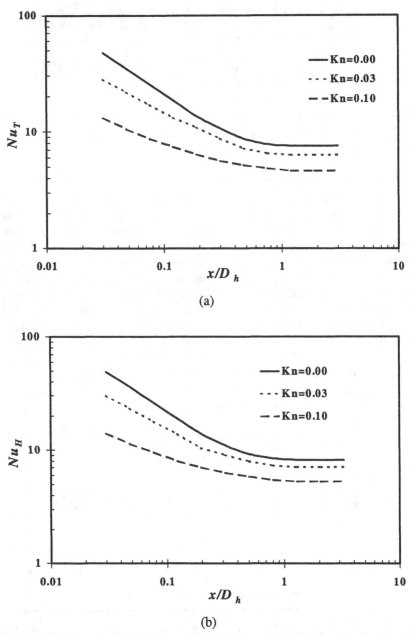

(a)

(b)

Figure 6.10: Numerical simulations of the slip flow effect, Knudsen number, on the local Nusselt number along a microchannel for: (a) uniform wall temperature, and (b) uniform heat flux boundary conditions [90].

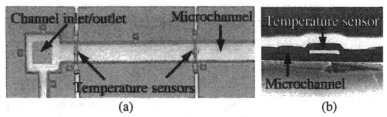

(a)   (b)

Figure 6.11: Photographs of: (a) a microchannel, about 1.4μm high and 40μm wide, integrated with suspended temperature microsensors, and (b) the channel cross-section showing a suspended microsensor, about 0.4μm thick.

A microchannel, integrated with suspended temperature sensors (Figure 6.11), has been constructed for an initial attempt to experimentally assess the slip flow effects on heat transfer in microchannels [80]. Nitrogen at room temperature is the working fluid, and the sensors measure the gas flow temperature. The measured distributions along the microchannel are shown in Figure 6.12 for a few constant wall-temperature boundary conditions. In all cases, the gas flow temperature along the channel is almost uniform and equal to the wall temperature; no cooling effect has been observed. Indeed, the slip flow effects are so small that the sensitivity of the experimental system is not sufficient to detect the temperature changes. Thus, experiments with higher resolution and greater sensitivity are required to accurately verify the weak slip flow effects on the temperature and the heat-transfer coefficient predicted by theoretical analyses and numerical simulations.

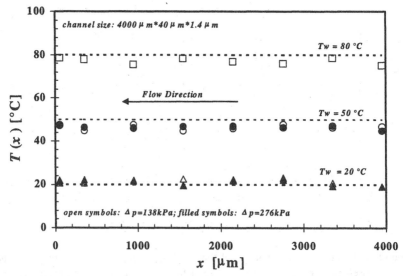

Figure 6.12: Measured streamwise temperature distributions of nitrogen flow in a microchannel for three constant wall temperatures under two total pressure drops.

6.5   Fully developed liquid flow forced convection

Liquid flow is considered to be incompressible even in micro ducts, since the distance between the molecules is much smaller than the characteristic scale of the flow. Hence, no rarefaction effect is encountered, and the classical model in Equation (6.8) should be valid. Again, in such a case, extensive data are readily available [182]. However, two unique features of liquid flow in micro ducts, polarity and electrokinetics, could affect the flow behaviour.

6.5.1   Electrokinetic effect (EDL)

The characteristic length scale of the electric double layer (EDL) is inversely proportional to the square root of the ion concentration in the liquid. For example, the EDL length scale is about 1μm in pure water, while it is only 0.3nm in 1 mole of NaCl. When a liquid is forced through a microchannel under hydrostatic pressure, the ions in the mobile region of the EDL are carried towards one end. This causes an electric streaming current in the direction of the liquid flow. The accumulation of ions downstream sets up an electric field with an electrical streaming potential. The streaming potential results in a conduction current flowing back in the opposite direction to the flow. A steady-state liquid flow develops when the conduction current equals the streaming current. The motion of the ions in the diffuse mobile layer affects the bulk of the liquid flow via momentum transfer due to viscosity. However, the ions motion is subject to the electrical potential in the diffuse layer. Thus, the liquid flow and its associated heat transfer may be affected by the presence of the EDL.

In macroscale flows, the interfacial electrokinetic effects are negligible, since the thickness of the EDL is very small compared to the hydraulic diameter of the duct. However, in microscale flow the EDL thickness is comparable to the hydraulic diameter, and the EDL effects must be considered in the analysis of fluid flow and heat transfer. The x-momentum and energy equations for a 2-D channel flow can be reduced to [122]:

$$\mu \frac{d^2 u}{dy^2} - \frac{dP}{dx} - \varepsilon\varepsilon_0 \frac{E_s}{L} \frac{d^2\psi}{dy^2} = 0 \qquad (6.20)$$

$$\rho c_p \left(u \frac{\partial T}{\partial x} \right) = k \left(\frac{\partial^2 T}{\partial y^2} + \frac{\partial^2 T}{\partial x^2} \right) + \mu \left(\frac{\partial u}{\partial y} \right)^2 \qquad (6.21)$$

where E_s is the steaming potential, and L is the duct length. Equation (6.20) can be integrated directly twice to yield an analytical solution for the velocity field. The constants of integration can be found by employing the boundary conditions:

$$u = 0 \quad \text{and} \quad \psi = \zeta \quad @ \quad y = \pm H/2 \qquad (6.22)$$

and the velocity profile is given by:

$$u(y) = -\frac{H^2}{8\mu}\frac{dP}{dx}\left[1 - \left(\frac{y}{H/2}\right)^2\right] -$$

$$-\frac{E_s\zeta\varepsilon\varepsilon_0 H^2}{4\mu L}\left[1 - \left|\frac{\sinh(y/\lambda_D)}{\sinh De}\right|\right] \qquad (6.23)$$

where H, ζ and λ_D are the channel height, zeta potential and Debye length, respectively. $De = H/2\lambda_D$ is the Debye number defined in Chapter 3. Using this velocity field, Equation (6.21) can be solved numerically for constant wall temperature boundary condition with a given inlet liquid temperature.

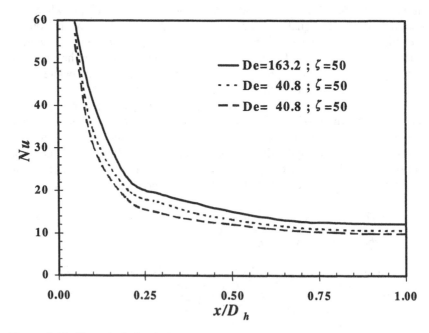

Figure 6.13: Numerical simulations of the electrokinetic effect, EDL, on the local Nusselt number along a microchannel with uniform wall temperature [122].

The results showed that both the temperature gradient at the wall and the difference between the wall and the bulk temperature decrease with downstream distance. The value of the temperature gradient decreases much faster resulting in a decreasing Nusselt number along the channel as evident in Figure 6.13. If there are no double layer effects, i.e. $\zeta=0$, a higher heat transfer rate (higher Nu) is obtained and, for the same value of De, Nu decreases as ζ increases. As De increases, e.g. weaker EDL field or smaller EDL thickness, the Nusselt number also increases. Thus, the EDL results in a reduced flow velocity (higher apparent viscosity), thus affecting the temperature distribution and leading to a smaller heat transfer rate.

6.5.2   Polarity effect

The theory of micropolar flow has received a great deal of attention in the literature. As the characteristic length scale of the flow approaches the molecular or the substructure size, even with electrically neutral solid boundaries, the fluid microstrucure has to be considered. In order to evaluate micropolar effects on the heat transfer in microchannels, Jacobi [74] considered the steady, fully-developed, laminar flow in a cylindrical microtube with uniform heat flux, for which the energy equation is given by:

$$\rho c_p \left(u \frac{\partial T}{\partial x} \right) = \frac{k}{r} \frac{\partial}{\partial r} \left(r \frac{\partial T}{\partial r} \right) \tag{6.24}$$

where r is the radial coordinate. The streamwise velocity $u(r)$ and the micro-rotation radial distribution $\omega(r)$ in a tube of radius R have been evaluated analytically as follows:

$$\frac{u(r)}{u_c} = 1 - \frac{r^2}{R^2} + \frac{\kappa}{\mu+\kappa} \frac{I_0(Er)}{Er \cdot I_1(Er)} \left[\frac{I_0(Er \cdot r/R)}{I_0(Er)} - 1 \right] \tag{6.25}$$

$$\frac{\omega(r)}{u_c / R} = \frac{r}{R} - \frac{I_1(Er \cdot r/R)}{I_1(Er)} \tag{6.26}$$

where u_c is the centreline velocity in classical Poiseuille flow. I_n is the modified Bessel function of the first kind of order n; μ, κ and γ are the viscosity coefficients for micropolar fluid. $Er=2R/l$ is the Eringen number defined in Chapter 3. By substituting the velocity field into the energy equation, integrating, and applying the boundary conditions: (i) the temperature is bounded for $r=0$, and (ii) the temperature is equal to the wall temperature at $r=R$, the radial temperature distribution $T(r)$ is calculated as:

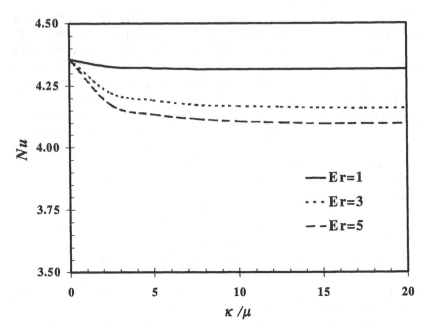

Figure 6.14: Analytical evaluation of the micropolar flow effect on the local Nusselt number along a microchannel with uniform heat flux [74].

$$\frac{T_w - T(r)}{q'' R / k} =$$

$$\left\{ \frac{3}{8} - \frac{r^2}{2R^2} + \frac{r^4}{8R^4} + \frac{\kappa}{\mu + \kappa} \frac{2}{Er} \frac{I_0(Er)}{I_1(Er)} \left[\frac{r^2}{4R^2} - \frac{1}{4} + \frac{1}{Er^2} \left(1 - \frac{I_0(Er \cdot r / R)}{I_0(Er)} \right) \right] \right\}$$

$$\left\{ \frac{1}{2} + \frac{\kappa}{\mu + \kappa} \frac{2}{Er} \left[\frac{1}{Er} - \frac{I_0(Er)}{2I_1(Er)} \right] \right\}^{-1} \qquad\qquad (6.27)$$

where q'' is the uniform heat flux. The deviation from the classical solution is determined by Er and κ/μ, and the model reduces to the classical Poiseuille flow for either $Er \to 0$ or $\kappa/\mu \to 0$. Based on the temperature field, the heat transfer rate can be calculated and the results are shown in Figure 6.14 for different values of Er. The Nusselt number is smaller than the classical value of $Nu=4.3636$. As $\kappa/\mu \to \infty$, Nu asymptotically approaches a lower value that depends on Er. The dependence of Nu on Er is not monotonic, and the minimum heat transfer rate of $Nu=4.0757$ is obtained for $Er=6.2$ and $\kappa/\mu=100$. Thus, micropolar flow effects also result in a reduced heat transfer rate.

6.5.3   Experimental results

The size effects in microchannel liquid flow have so far been discussed based on theoretical considerations only. The pioneering work of Tuckerman and Pease [208] included measurements of heat transfer and pressure drop in microchannels, about 50μm wide and 300μm deep, using deionised water as the working fluid. The total thermal resistance was found to be independent of the mass flux, as expected for fully-developed laminar flow. Peng and Peterson [149] studied convective heat transfer and flow friction for water flow in microchannels with hydraulic diameter in the range 100-400μm, with the Reynolds numbers ranging from 50 to 4000. For both the laminar and the turbulent regimes, the Nusselt number was found to be a function not only of the Reynolds and Prandtl numbers but also of geometric parameters. For laminar flow the following correlation is suggested:

$$Nu = 0.1165 \left(\frac{D_h}{W_c} \right)^{0.81} \left(\frac{H}{W} \right)^{-0.79} Re^{0.62} Pr^{1/3} \tag{6.28}$$

For the turbulent flow, the suggested correlation is:

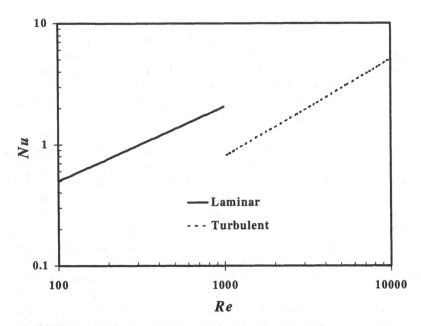

Figure 6.15: Nusselt number dependence on the Reynolds number for water flow in microchannels under laminar and turbulent flow conditions.

$$Nu = 0.072 \left(\frac{D_h}{W_c} \right)^{1.15} \left[1 - 2.421 \left(\frac{H}{W} - 0.5 \right)^2 \right] Re^{0.8} Pr^{1/3} \qquad (6.29)$$

where W_c is the center-to-center distance between channels, D_h is the hydraulic diameter, H and W are the channel height and width, respectively. The Nusselt number is plotted in Figure 6.15 as a function of the Reynolds number for a square channel, $H/W=1$, with $D_h/W_c=0.1$ and $Pr=4$.

Tso and Mahulikar [205-207] have presented several articles advocating the inclusion of the Brinkman number in the heat transfer correlation for laminar flow in microchannels. The Brinkman number, $Br=\mu \cdot u_a/k \cdot \Delta T$, reflects the relative importance of viscous heating to fluid conduction. As viscous heating is negligible in macroscale flows, the Brinkman number is not included in the classical heat transfer correlations. However, velocity gradients in microchannels are expected to be high with large length-to-diameter ratio. Thus, utilizing dimensional analysis, the following correlation is suggested:

$$Nu = A \cdot Re^a \cdot Pr^b \cdot Br^c \cdot \left(\frac{L}{D_h} \right)^d \qquad (6.30)$$

where L is a critical dimension of the microchannel. The exponent of Br, c, is positive for heating and negative for cooling of the fluid while the exponents used for Re and Pr, a and b, are the same as in Equation (6.28). The reported values of Br for water flow are on the order of 10^{-8}-10^{-7}, indicating that even for microchannel flow ($D_h \sim 700 \mu m$) viscous dissipation is negligible compared with the liquid heat conduction.

Clearly, the electrokinetic and micropolar flow effects on liquid forced convection in micro ducts are indirect. Namely, the velocity is modified due to these effects and, as a consequence, the heat transfer rate is affected. Thus, it is important to first verify the hydrodynamic effects. Indeed, it has been suggested in a few reports that theoretical calculations based on the classical model did not agree with experimental measurements of liquid flow properties in microchannels [149,166]. An experimental study of water flow in a microchannel with a cross-section area of $600 \mu m \times 30 \mu m$ was carried out by Papautsky *et al.* to specifically evaluate micropolar effects [141]. They concluded that micropolar fluid theory provides a better approximation to the experimental data than the classical theory. However, a close examination of the results shows that the difference between the results of the two theories is smaller than the difference between the experimental data and the predictions of either theory.

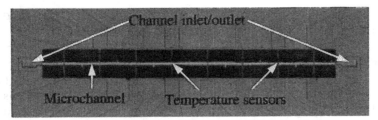

Figure 6.16: A composite picture of a suspended microchannel integrated with temperature sensors on the channel roof.

In carefully conducted experiments of water flow through a suspended microchannel with a cross section area of about 20μm×2μm, Figure 6.16, none of these effects have been observed [228]. The measured mass flow rate as a function of the pressure drop is compared in Figure 6.17 with the calculated rate based on the classical plane Poiseuille flow given by:

$$Q_m = -\rho \frac{WH^3}{12\mu} \frac{\partial P}{\partial x} \qquad (6.31)$$

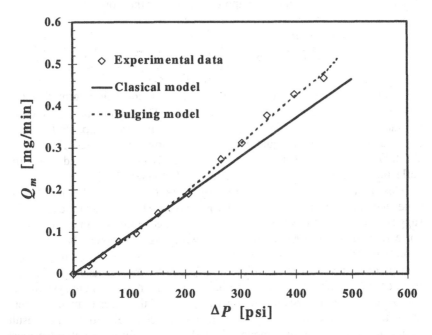

Figure 6.17: A comparison between mass flow rate measurements and calculations as a function of the pressure drop. The calculations are based on both the classical and the bulging models for water flow in the microchannel shown in Figure 6.16.

where H, W and L are respectively the channel height, width and length; ρ, μ and ΔP are the liquid density, viscosity and total pressure drop along the channel, respectively. Evidently, the measured flow rate increases non-linearly with the pressure drop, in contrast with the analytical model for the incompressible water flow. The measured flow rate is higher, suggesting that the channel deforms due to the high internal pressure. This hypothesis is tested by modifying the channel height in the volume flow rate calculations, accounting for the top-surface deflection due to pressure, such that:

$$H = H_0[1 + a \cdot P(x)] \qquad (6.32)$$

where H_0 is the channel nominal height, and a is a constant. Substituting this modified height into Equation 6.31, with the assumption $a \cdot P(x) \ll 1$, yields:

$$Q_m = -\rho \frac{WH_0^3}{12\mu}[1 + 3aP(x)]\frac{\partial P}{\partial x} \qquad (6.33)$$

Integration along the channel, assuming $P_i - P_o \gg P_o$ due to the high inlet pressure, results in the following approximation for the flow rate:

$$Q_m = -\rho \frac{WH_0^3 \Delta P}{12\mu L}\left(1 + \frac{3}{2}a \cdot \Delta P\right) \qquad (6.34)$$

By comparing Equations (6.31) and (6.34), the correction term due the bulging effect, $3a\Delta P/2$, can be identified. This geometric correction introduces a parabolic dependence on the pressure drop. Fitting the experimental data with Equation (6.34), as shown in Figure 6.17, yields $a = 0.000132\text{psi}^{-1}$. The channel bulging has visually been confirmed by in-situ deflection measurements using an optical profilometer [224].

This is a particularly revealing study demonstrating the pitfalls of experimental measurements in microsystems. All the theoretical works suggest that the known size effects described hitherto may introduce corrections of up to 10% in comparison with classical models based on the continuum assumption. However, in many instances, it is difficult to obtain measured values with such accuracy, especially when measuring the heat transfer rate. At times the errors are inherent in the measurement system, i.e. sensor resolution and accuracy. However, even if the measurements system is sufficiently accurate, other effects such as the channel bulging may result in deviations from classical models. The channel height is probably the most important geometric parameter, and in-depth error analysis of the microchannel height has recently been presented [240].

Chapter 7

Steady, Forced Convection Boiling in Micro Ducts

Convective boiling is defined as the addition of heat to a flowing liquid in such a way that generation of vapour occurs while, conversely, convective condensation is defined as the removal of heat from the fluid in such a way that vapour is converted into liquid. This definition, therefore, excludes the process of flashing where vapour generation occurs solely as the result of a reduction in flow pressure. However, in many systems, the two processes do occur simultaneously and, hence, cannot be clearly distinguished. This chapter will be concerned with only single-component microsystems, i.e. a pure liquid and its vapour. Furthermore, much of the information presented will be devoted to one such system, namely the water/steam system. Many other fluid systems of industrial importance, however, involve the use of multi-component systems such as refrigerants, organic liquids, cryogenic liquids and liquid metals. Very limited information is available on the processes of boiling and condensation with multi-component macrosystems and practically none with multi-component microsystems.

As devices become more compact and powerful, removal of excessive heat from tiny regions with reduced operating temperatures emerges as a formidable challenge. Under these extreme conditions, there are few alternatives to boiling heat transfer. The efficacy of boiling comes in the agitation of the near wall fluid by departing bubbles, which grow within pits in the heated solid surface. The agitation occurs where it is needed most, i.e. in the laminar layer adjacent to the boundary where turbulence would be damped by the presence of the wall. Moreover, as the heat flux rises, the boiling activity increases with more sites becoming active, and each site emitting bubbles at an increasing frequency. The result is a mechanism by which there is a minimal increase in wall temperature with great increase in convective heat flux. This, indeed, represents an ideal mode of heat transfer.

Two problems associated with forced convection boiling, however, have prevented its implementation particularly in microsystems. One is the incipience to boiling or onset of nucleate boiling (ONB) and the other is the critical heat flux (CHF). These two events will be discussed along with the evolving flow pattern and thermal response due to the rising input heat flux.

7.1   Liquid-to-vapour phase change

It has been recognized from studies of phase change that vapour may form in several ways corresponding to the departure from an equilibrium state [29]. The formation of vapour at a planar interface occurs when the liquid temperature is increased fractionally above the corresponding saturation temperature. The term evaporation will be reserved for this process denoting vapour formation at a continuous liquid surface such as the interface between the liquid film and the vapour core in annular flow. Evaporation and condensation at a planar interface can be described in terms of the imbalance of molecular fluxes passing through the interface from the vapour and liquid phases respectively. The vaporization acts as a heat sink at the surface, and heat transfer coefficients from the heating wall to the liquid can be formed with the driving temperature difference $\Delta T_{sh}=T_w-T_s$; T_w is the temperature of the superheated wall surface, while T_s is the liquid saturation temperature. The laws of free convection heat transfer apply for evaporation due to wall heat flux q, i.e. $q=h\cdot\Delta T_{sh}$. Thus, for laminar flow [71]:

$$h \sim \Delta T_{sh}^{1/4} \quad \text{or} \quad h \sim q^{1/5} \tag{7.1-a}$$

and for turbulent flow:

$$h \sim \Delta T_{sh}^{1/3} \quad \text{or} \quad h \sim q^{1/4} \tag{7.1-b}$$

Boiling is a phase change process in which vapour bubbles are formed either on a heated surface or in a superheated liquid layer adjacent to the heated surface. It differs from evaporation at predetermined vapour/liquid interfaces because it also involves creation of these interfaces at discrete sites. In a metastable liquid, there is a small but finite probability of a cluster of molecules coming together to form a vapour embryo of the size of the equilibrium nucleus. Homogeneous vapour nucleation will occur if further molecules collide with an equilibrium embryo. Although homogeneous nucleation does occur in organic liquids, it can be discounted as a mechanism for vapour formation at least in water. On the other hand, foreign bodies and solid boundaries provide ample nuclei to act as centres of vapour formation. This mechanism of vapour generation from pre-existing nuclei is termed heterogeneous nucleation. Nucleate boiling is a very efficient mode of heat transfer, and the heat transfer coefficient is approximately proportional to the third power of the temperature difference [71]:

$$h \sim \Delta T_{sh}^3 \quad \text{or} \quad h \sim q^{3/4} \tag{7.2}$$

In convective boiling, the flow conditions are determined to a large extent by the pressure gradient along the heating surface. The vapour content increases along the path of flow up to the point of complete vaporization. A variety of boiling phenomena occur in correspondence to the liquid flow rate. In general, the working liquid enters subcooled into a heated duct. Vapour bubbles, formed at the wall, re-condense in the colder core of the liquid flow. If the liquid in the core is heated up to saturation temperature, nucleate boiling results where the heat transfer coefficient is chiefly determined by the heat flux. Under forced convection, the coefficient depends weakly upon the flow rate. The individual bubbles grow together into large bubbles, forming a succession of flow patterns with a continuous increase of the vapour content.

7.2   Nucleate boiling

Vapour/gas trapped in imperfections such as cavities and scratches on the heated surface serve as nuclei for bubbles. The density of active sites increases as wall heat flux or wall superheat increases. Clearly, addition of new nucleation sites influences the heat transfer rate from the solid surface to the working fluid. Several other parameters also affect the site density, including the surface finish, surface wettability, and material thermophysical properties. After inception, a bubble continues to grow until forces causing it to detach from the surface exceed those pushing the bubble against the wall. Bubble dynamics, which plays an important role in determining the heat transfer rate, includes the processes of bubble growth, bubble departure, and bubble release frequency [37].

7.2.1   Bubble formation

If the wall temperature rises with increasing heat flux, vapour bubbles will form at a certain nucleation sites when a definite wall superheat temperature is reached. The nucleation sites are gas- or vapour-filled cracks and cavities on the heated wall. A wedge-shaped imperfection on a surface will trap vapour/gas, by an advancing liquid front, if the contact angle is larger than the wedge angle, i.e. $\psi > \phi$, as illustrated in Figure 7.1 [7].

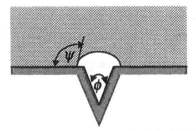

Figure 7.1: Bubble nucleation from a cavity with an included angle φ and a contact angle ψ between the gas/vapour and liquid.

According to a more refined criterion, based on minimization of the Helmholtz free energy of a system involving a liquid/gas interface, a cavity will trap vapour/gas if [212],

$$\psi > \varphi_{min} \qquad (7.3)$$

where φ_{min} is the minimum cavity-side angle of the cavity. While the first criterion is a necessary condition for vapour/gas entrapment in a wedge, Equation 7.3 provides a sufficient condition.

Several approaches have been proposed for determining the incipient wall superheat for boiling from pre-existing nuclei. The leading approach suggests that boiling incipience corresponds to a critical point of instability of the vapour-liquid interface. The interface is considered to be stable or quasi-stable if its curvature increases with an increase in vapour volume. An investigation of the vapour-liquid interface instability in a spherical cavity showed that nucleation occurs when the interface curvature attains a maximum value. A relation between wall superheat and the diameter of the nucleation cavity has been proposed as follows [212]:

$$\Delta T_{sh} = \frac{4\sigma T_s}{\rho_v h_{fg} D_c} K_{max} \qquad (7.4)$$

where $K_{max}=1$ for $\psi \leq 90°$ and $K_{max}=\sin\psi$ for $\psi>90°$; σ is the interfacial tension, ρ_v is the vapour density, h_{fg} is the fluid latent heat of vaporization, and D_c is the cavity-mouth diameter. The interface temperature is assumed to be equal to the wall temperature. According to this model, few pre-existing vapour/gas nuclei are found for well wetting liquids. Therefore, the expected wall superheat at nucleation should approach the homogeneous nucleation temperature. The observed inception superheats for these liquids, however, are much smaller than those corresponding to homogeneous nucleation. Gases dissolved in liquids, e.g. air, trigger early nucleation and, consequently, reduce the required superheat temperature. Indeed, in many applications, gas is added by external means into liquids in order to reduce the inception temperature.

The density of active nucleation sites increases as wall heat flux or superheat temperature increases; thus, enhancing the heat transfer rate from the solid surface. For water, the active site density N_a is correlated as [37]:

$$N_a \sim \Delta T_{sh}^m \qquad (7.5)$$

where m varies between 4 and 6. The proportionality constant in Equation 7.5 was reported to increase with surface roughness, but the exponent m was independent of surface roughness [33]. Thermal and flow conditions in the vicinity of a heated surface as well as interactions between neighbouring sites can lead to activation of inactive sites and deactivation of active sites. Thermal interactions are expected to alter the local active-site density and distribution only at low heat fluxes or in partial nucleate boiling. However, the significance of these processes with respect to heat transfer during well-developed nucleate boiling is expected to be negligible.

7.2.2   Classical bubble dynamics

Once the vapour nucleus has attained a size greater than that for unstable equilibrium it will grow spontaneously. The growth may be limited initially by the inertia of the surrounding liquid or, at a later stage, by the rate at which the latent heat of vaporization can be conducted to the vapour-liquid interface. Most of the evaporation occurs at the base of the bubble, where a microlayer between the vapour-liquid interface and the heated surface exists. The bubble growth is mainly due to evaporation from the microlayer and, thus, microlayer evaporation is a significant contributor to the heat transfer during bubble growth. The thickness of this microlayer is estimated as [30],

$$\delta \sim \sqrt{v_f t_g} \qquad (7.6)$$

where v_f is the liquid kinematic viscosity, and t_g is the bubble growth time. However, a consistent model for bubble growth on a heated surface that properly accounts for the microlayer contribution and time-varying temperature and flow field around the bubble is still lacking.

A vapour bubble grows to a certain diameter before departure, and the balance of forces acting on the bubble dictates this diameter. In general, during the later phase of bubble growth, buoyancy and hydrodynamic drag forces act to detach the bubble from the surface against surface tension and liquid inertia forces that pull the bubble toward the surface. In small cavities, the departure bubble size is mainly dictated by a balance between buoyancy and inertia forces. On the other hand, for larger cavities, the bubble size at departure is set by a balance between buoyancy and surface tension forces. A correlation for the bubble departure diameter is given by [29]:

$$D_A \sim 0.851\psi' \sqrt{\frac{2\sigma}{g(\rho_f - \rho_g)}} \qquad\qquad (7.7)$$

where ψ' is the contact angle in radians. Although significant deviations of the bubble diameter at departure with respect to Equation 7.7 have been reported, especially at high system pressure, this correlation does provide a correct length scale for the boiling process. More complete expressions have been reported for departure bubble diameter, obtained either empirically or analytically. However, these expressions are not consistent with respect to the role of the surface tension or the effect of the wall superheat on the bubble departure process [37].

Subcooled boiling is characterized by the formation of vapour at the heating surface in the form of bubbles at certain nucleation sites. These vapour bubbles condense in the cold liquid away from the heating surface; hence, vapour formation and condensation are present simultaneously. While the bubble is still growing at the base, condensation is already occurring at the top. Maximum bubble diameter and condensation time decrease with increasing subcooling temperature $\Delta T_{sc}=T_s-T_f$, where T_f is the bulk liquid temperature. Also, an increasing flow velocity impedes bubble growth. At high heat fluxes the onset of subcooled boiling is encountered at high degrees of subcooling, and the vapour bubbles may grow and collapse whilst still attached to, or even sliding along, the heating surface, since the condensation takes place much more rapidly than the growth of the bubbles. Hence, the maximum volume and life span of the bubbles are very small.

Individual nucleation sites emit bubbles with a certain mean frequency, which varies from site to site. The bubble departure frequency f is equal to the reciprocal of the sum of the bubble waiting time t_w and growth time t_g. The waiting time corresponds to the time it takes for the development of the thermal layer to allow nucleation of a bubble. Predictions of bubble release frequency based on waiting and growth time, however, do not match well with the data because many simplifications are made in calculating t_w and t_g. Initial correlations were based on the approximation $f \cdot D_a = const$. Others suggested correlations of the form $f \cdot D_a^n = const$, with $n=1/2$. More precise investigations have shown that the exponent n is not constant, but rather ranges between 0.5 and 2 [190].

7.2.3   Bubble activity in micro ducts

It has been reported in a few studies that, under certain circumstances, there must be a considerable superheat, i.e. rise of wall temperature above the fluid saturation temperature, before boiling can begin in microchannels. The temperature level required to initiate boiling may then be larger than the allowable maximum wall temperature. Thus, depending on the operation, nucleate boiling state may not be reached without sustaining device damage while in the liquid single-phase regime. On the other hand, others reported that nucleate boiling in micro-channels is greatly intensified, and the wall superheat for flow boiling may be much smaller than that for macrochannels under the same heat flux. Furthermore, some reported that no partial nucleate boiling of subcooled water through microchannels was observed, and no bubbles were detected in the nucleate boiling regime [154].

In a microchannel heat sink, local bubble nucleation was experimentally observed within triangular microchannels with a hydraulic diameter of 26μm [85]. The bubble activity started at input power level as low as $q/q_{CHF}\cong0.5$, where q_{CHF} is the power input required for the onset of critical heat flux condition. The working fluid was water and the corresponding average device temperature was about 70°C. Bubbles could be seen forming at specific locations along the channel walls at very few active nucleation sites. The bubble release frequency inside the micro ducts was fairly high, and a picture of a mature bubble is shown in Figure 7.2. However, there were very little, if any, active nucleation sites along the channel walls. Furthermore, most of the nucleation sites became inactive after one or two tests suggesting that they may have been residues of the fabrication process. Therefore, no attempt was made to characterize the bubble release frequency.

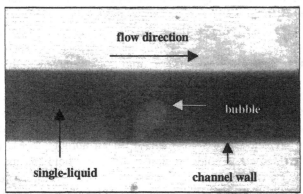

Figure 7.2: An active nucleation site within a microchannel exhibiting a cyclic sequence of bubble formation, growth and departure (50μm width, array of 35 microchannels, q/q_{CHF}=0.4) [85].

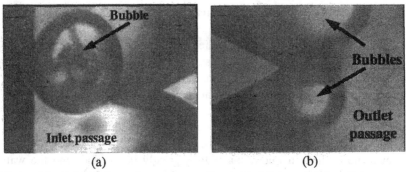

(a)   (b)

Figure 7.3: Bubble formation and growth at (a) inlet, and (b) outlet common passages of a microchannel array (passages are about 100μm wide and 40μm deep).

At slightly higher input power level, $0.5<q/q_{CHF}<0.6$, large bubbles were generated at the inlet/outlet common passages connecting the microchannel array to the device common inlet/outlet. The boiling activity at these larger passages, shown in Figure 7.3, became more intense with increasing input power; namely more active sites with increasing release frequency. The upstream bubbles, typically larger than the duct cross-section, were forced through the micro ducts. These bubbles, upon departure from their nucleation sites, blocked the duct entrance until the upstream pressure was high enough to force them into the micro duct. In some cases, the bubbles travelled slowly along the channel as slug flow. In most instances, however, the bubbles were ejected at high speed through the micro duct and could not be detected until they re-appeared at the channel exit.

The channel triangular cross-section, though convenient to form in (100) Si wafer using TMAH, resulted in poor flow visualization due to the angled reflection of the incident light from the sidewalls. The problem has been alleviated in a similar device with nearly rectangular microchannels [107]. The experiments re-confirmed that as long as a nucleation site is active, a clear formation of a bubble, its growth and departure is observed inside a microchannel. Regardless whether the origin of the bubble is dissolved gas or liquid phase change, this cycle of bubble activity has been observed in a 14μm-deep microchannel as documented in the picture series of Figure 7.4.

An interesting feature, to which the channel small size probably contributes, is the bubble evolution after its departure. In the classical process, the vapour bubbles re-condense in an abrupt process (collapse); thus, transferring heat to colder regions in the flow field and maintaining a temperature field with very small gradients. However, in the present case, the bubbles did not re-condense but, as viewed by a regular-speed camera in the pictures shown in Figure 7.5, seemed to travel undetected at a very high speed along the channel.

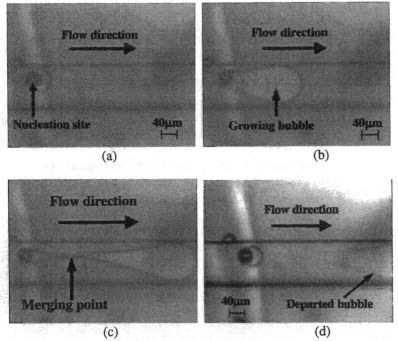

Figure 7.4: Bubble activity at an active nucleation site: (a) bubble formation, (b) growth, (c) departure, and (d) departed bubble next to newly formed bubble (channel average width is 120μm and depth 14μm) [107].

The bubbles disappeared immediately following their departure from the nucleation site, Figure 7.5-a. The bubbles then re-formed, after a certain time delay, upon impingement on the channel exit wall, Figure 7.5-b. Subsequently, the bubbles continued to move toward the device single outlet and out of the system. The bubble motion seems to be adiabatic with no heat transfer to the surrounding liquid, since the re-formed bubble size is about the size of the detached bubble. These observations clearly indicate that the bubbles had originated from dissolved gases in the DI water since the device temperature, about 60°C, was far below the water saturation. Consequently, the departing bubbles did not condense back to liquid phase, but rather reformed after impact at the microchannel exit.

The physical mechanism of bubble departure in a micro domain could be different from classical models. Figure 7.4-c suggests that the bubble is detached when the two vapour/liquid interfaces merge together along the stem between the bubble and the nucleation site. The bubble is snapped off from the nucleation site, and is free to move within the fluid flow. This is a direct result of the channel small size, since the bubble growth is limited by the channel solid walls and not by surface tension or vapour condensation.

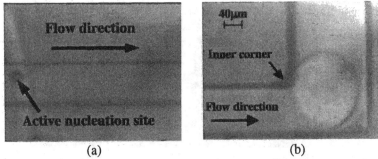

(a)   (b)

Figure 7.5: Bubble motion through the microchannel: (a) bubble departure from its nucleation site, and (b) bubble re-formation upon impact on the channel exit wall (channel average width is 120μm and depth 14μm) [107].

Flow visualizations using a high-speed camera reveal a sequence of flow patterns that were not detected by a regular camera. Bubble formation, growth and departure are rather unique when the wall temperature is lower than the water saturation temperature. The active nucleation sites are always at defects, such as pinholes formed on the channel walls during etching. In subcooled nucleate boiling, vapour bubbles grow to a critical size prior to their departure and condense in the cold liquid. However, due to the limited space in a microchannel, the bubble grows up until it occupies the entire cross-section. Resting against the solid walls, the bubble is then stretched in the flow direction. A very long stem connects the bubble to its nucleation site as shown in Figure 7.6-a. Only when the surface tension along the stem overcomes the internal pressure such that both interfaces merge, does the bubble depart. The bubble then loses its distinct boundary, as indicated in Figure 7.6-b, but it does not condense; instead the bubble re-forms upon impact at the channel exit indicating that it originated from dissolved gas.

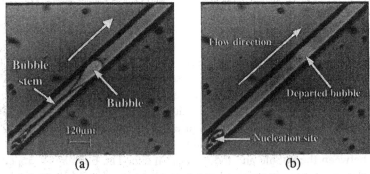

(a)   (b)

Figure 7.6: High-speed camera pictures of: (a) mature bubble with a stretched stem, and (b) departed bubble after the merging of the liquid/gas interfaces along the stem (channel average width is 120μm and depth 14μm) [108].

Further increase of the input power level, $q/q_{CHF}>0.7$, resulted in an annular flow pattern, and bubble nucleation on the duct walls could no longer be observed. The corresponding device temperature was about 90°C. It seems very likely that the suppression of the nucleation sites within the micro duct was the result of the evolution of a more stable flow pattern rather than a genuine size effect. It is reasonable to expect the bubble dynamics after inception, unlike the nucleation site density, to be affected by the channel size. Still it has been demonstrated that in microchannels, with hydraulic diameter as small as 25μm, bubble growth and departure is possible. Thus, the lack of partial nucleate boiling of subcooled liquid flowing through microchannels cannot be attributed to a direct fundamental size effect suppressing bubble dynamics, i.e. bubbles cannot grow and depart due the small size of the channel. However, it is very plausible that the absence of partial nucleate boiling is an indirect size effect. Namely, another boiling mode such as annular flow becomes dominant due to small channel size, i.e. strong capillary forces, and as a result the bubble dynamics mechanisms are suppressed. Naturally, due to the fabrication process and the channel small size, the interior surfaces are very smooth. Therefore, very few active nucleation sites can be observed, almost always at defects on the channel walls due to the fabrication process (e.g., pinholes during etching).

7.3   Flow patterns

Two-phase flow patterns in ducts are the result of the detailed heat transfer between the solid boundary and the working fluid. The flow patterns are important since the temperature distributions in both the solid boundary and the fluid flow are directly determined by these patterns. Various hydrodynamic conditions encounter when a duct is heated uniformly over its length and fed with subcooled liquid at its entrance at such a rate that the liquid is totally evaporated at duct exit. Although gravity is important in macrosystems, the flow pattern evolution in convective boiling as sketched in Figure 7.7 is similar whether in a vertical or a horizontal duct. Whilst the liquid is being heated up to the saturation temperature and the wall remains below the superheat temperature necessary for nucleation, the process is single-phase convective heat transfer from the wall to the working liquid. At a certain point, the conditions adjacent to the wall allow the formation of vapour bubbles from nucleation sites giving rise to bubbly flow and, further downstream, single bubbles merge into larger ones resulting in plug or slug flow. The heat transfer in this region is dominated by subcooled or saturated nucleate boiling. With increasing vapour content, a point is reached where a fundamental transition in the heat transfer mechanism from boiling to evaporation occurs corresponding to a transition to annular flow.

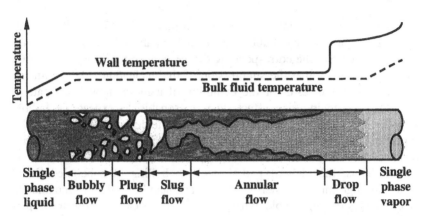

Figure 7.7: Wall and fluid temperature distribution with respect to the flow patterns in a uniformly heated duct [190].

In the annular flow region, the thickness of the liquid film on the surface is thin enough such that the effective thermal conductivity is sufficient to prevent the liquid in contact with the wall being superheated to the temperature required for bubble nucleation. Heat is carried away from the wall by forced convection in the film to the liquid-vapour core interface, where evaporation takes place. Since nucleation is completely suppressed, the heat transfer can no longer be called boiling, and the heat transfer is referred to as two-phase forced convection. The liquid film thickness in the annular flow region continuously decreases until the liquid is completely evaporated at a certain position resulting in drop flow of vapour with liquid droplets. This dryout transition is accompanied by a rise in the wall temperature for ducts operating under a controlled surface heat flux. Finally, single-phase convective heat transfer to vapour develops after the transition of the working fluid to dry saturated vapour near the duct exit.

Mudawar and Bowers [134] reported that low and high mass velocity flows in small diameter tubes, 0.4-2.5mm, are characterized by drastically different flow patterns as well as unique CHF trigger mechanisms. The combination of low flow velocity and a long tube often results in a saturated vapour flow at the tube exit while, at the entrance, heat is transferred to the subcooled liquid by single-phase forced convection. In between, the flow exhibits a succession of bubbly, slug and annular flow. The high velocity flow is characterized, as in the low velocity, by an inlet region of single-phase convection until the onset of nucleate boiling. Once nucleation begins, very small bubbles are formed creating a thin bubble boundary layer, and bubbles migrating toward the core quickly condense. The flow remains in the subcooled boiling region over its entire heated length. It is not clear whether this distinction is valid for microchannel heat sinks as well.

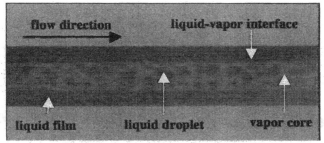

Figure 7.8: Unstable annular flow mode in a triangular microchannel with liquid droplets in the vapour core (q/q_{CHF}=0.6, 50μm width, 35 microchannels).

An integrated microsystem, similar to the one shown in Figure 4.4, has been fabricated to study the forced convection boiling flow patterns [85]. The triangular-shaped grooves etched in the silicon wafer were covered by a bonded glass wafer in order to facilitate flow visualizations. In micro ducts, body forces such as gravity are negligible with respect to surface forces, i.e. capillary or friction forces. Consequently, the micro duct orientation has little effect on forced convection boiling, and no difference between the flow patterns in horizontal and vertical micro ducts could experimentally be detected. Furthermore, the boiling modes identified in these micro ducts are different from the classical patterns sketched in Figure 7.7. At moderate power levels, an annular flow mode with liquid droplets within the vapour core could be observed, as shown in Figure 7.8, while the interface between the liquid film and the vapour core appears to be wavy. This mode should be regarded as an unstable, transition stage, since it was not always detected. Moreover, when it did appear, it was short lived. An annular flow mode, shown in Figure 7.9, was observed to be a stable pattern for a wide range of input power level, 0.6<q/q_{CHF}<0.9. An interface between the liquid film and the vapour core was clearly distinguishable. No liquid droplets existed within the vapour core, indicating that the vapour-core temperature was higher than the liquid saturation temperature.

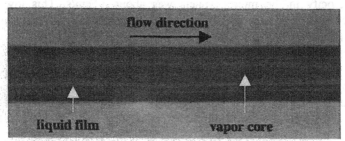

Figure 7.9: Stable annular flow mode in a microchannel with sharp interfaces between the liquid and vapour regions (q/q_{CHF}=0.8, 50μm width, 35 microchannels).

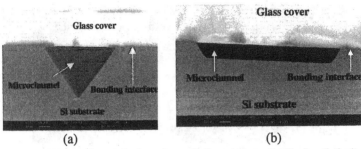

Figure 7.10: SEM pictures of microchannels between anodically bonded glass and silicon substrates with: (a) triangular and (b) nearly rectangular cross-section.

Evaporation at the liquid film-vapour core interface dominated the heat transfer from the channel wall to the fluid in the annular flow mode. Since the heat is conducted through the liquid film to the interface, the temperature at the wall has to increase to allow higher heat transfer rate enforced by the increased input power. The temperature would increase linearly with the input power if the film thickness stays constant. However, the film thickness decreased with increased power, due to the evaporation process, resulting in higher void fraction of the two-phase exit flow. Thus, the input power is converted into: (i) latent heat required for the evaporation at the liquid-vapour interface due to the phase change; (ii) internal energy of the liquid film manifested by the increased liquid and wall temperature. The combination of the two mechanisms resulted in monotonic temperature increase with decreasing slope as the input power increased.

It is not clear whether the stable annular flow is a general pattern in microchannels due to size effect, or it is unique only to the triangular channel cross-section shown in Figure 7.10-a due to the strong capillary forces at the sharp corners (similar to micro heat pipes). Therefore, an identical device with the exception of the nearly rectangular cross-section shown in Figure 7.10-b was fabricated to test the possible channel shape effect on the evolving flow patterns. At a low water flow rate, the average flow velocity in the microchannels was about 3.5cm/s. This low speed provides sufficient time for the fluid to exchange heat with the ambient through the 'hot' silicon and 'cold' glass substrates, resulting in vapour condensation on the glass inner surface. Furthermore, the large aspect ratio of the channels rectangular cross-section, about 8.5, resulted in a large surface area for free convection heat transfer. Thus, the combination of low flow velocity and large free convection area allowed a significant portion of the heat to be conducted through the device and transferred to the surrounding. The loss of heat is accompanied by re-condensation of the vapour phase forming liquid droplets on the colder glass ceiling, as shown in Figure 7.11-a, even at the channel inlet next to the heat source.

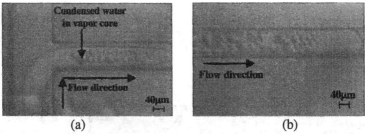

(a)   (b)

Figure 7.11: A picture showing the two-phase flow pattern at: (a) the channel entrance, next to the heater (T=86°C), and (b) next to the channel exit [107].

Since the heater was located near the inlet, the channel wall temperature decreased towards the heat sink outlet. Consequently, a transition region between the upstream vapour and the downstream liquid was established as shown in Figure 7.11-b. Close to the onset of CHF condition, the transition region shifted downstream towards the channel exit. Clearly, the vapour flow at the upstream part of the channel suppressed the bubbly flow mode as in the triangular microchannels. However, in this case, the flow void fraction increases with increasing power not only due to the streamwise motion of the transition region but also due to re-vaporization of the liquid droplets. At higher water velocity with smaller free convection area, vapour re-condensation did not occur as no liquid droplets were observed in the vapour core. No annular flow was observed in the rectangular channel under a CCD camera. However, using a high-speed camera, 500 frames per second, an annular flow field was detected in the microchannel as demonstrated in Figure 7.12. This flow mode was highly unstable and, within 0.3sec of its formation (Figure 7.12-a), the liquid films on each side of the vapour core were attracted to each other due to surface tension. After the merger of the two liquid films (Figure 7.12-b), the vapour core snapped immediately and the annular flow mode disappeared (Figure 7.12-c). This is different from the triangular microchannels, where the annular flow was the most stable mode.

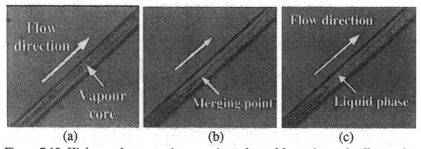

(a)   (b)   (c)

Figure 7.12: High-speed camera picture series, a-b-c, of formation and collapse of an unstable annular flow in a nearly rectangular microchannel 120μm in width [108].

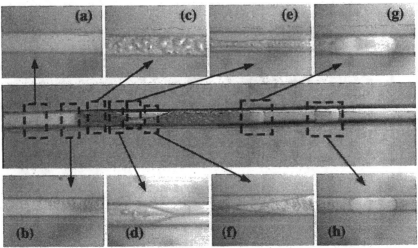

Figure 7.13: A sequence of pictures, a-b-c-d-e-f-g-h, detailing the transition region from the upstream vapour (a) to the downstream liquid zone (h) in two-phase convective boiling flow. Flow direction is from left to right [108].

The most prevalent boiling flow pattern observed in a rectangular microchannel was a transition region separating an upstream vapour zone from a downstream intermittent zone. The transition region seemed to oscillate randomly along the microchannel with no clear structure. However, under the CCD camera, a fascinating orderly pattern was observed as detailed in Figure 7.13. Upstream of the transition region (Figure 7.13-a), the flow was in single vapour phase since the heater was located at the channel inlet. At the upstream tail of the transition region (Figure 7.13-b), liquid droplets started to condense on the colder channel glass ceiling. The liquid droplets increased in number and size (Figure 7.13-c), until two liquid films were condensed on both sidewalls (Figure 7.13-d). Thus, an annular flow pattern was formed over a short segment (Figure 7.13-e). The liquid films became wider until they touched each other (Figure 7.13-f). When this occurred, a vapour section at the head of the transition region was severed from the upstream vapour zone (Figure 7.13-g). Finally, after the two liquid films merge together, the upstream vapour region was completely snapped off (Figure 7.13-h), leaving a vapour segment bounded by liquid flow. The entire transition region retreated far upstream, and the cycle repeated itself resulting in an intermittent flow downstream of the transition region. The transition region oscillated violently along the microchannel with large amplitude, about 500μm, and its average location depended on the input power. As the heat flux increased, the transition region moved downstream appearing at the channel exit when the input power reached the CHF level.

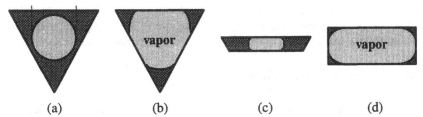

(a)   (b)   (c)   (d)

Figure 7.14: Schematic illustrations of an annular flow mode in: (a) a triangular cross-section with a continuous liquid film, (b) a triangular cross-section with contact lines, (c) a shallow rectangular cross-section with contact lines across the channel, and (d) a deep rectangular cross-section with contact lines at the corners.

Apparently the channel cross-section geometry does have an effect on the evolving flow pattern due to convective flow boiling in microchannels. Annular flow mode typically includes a continuous thin liquid film coating the inner channel surface with no contact lines as illustrated in figure 7.14-a. The sharp interfaces on the triangular channel ceiling suggest the existence of contact lines as sketched in Figure 7.14-b. This flow pattern appears to be very stable in the triangular cross-section. If the vapour pressure decreases, the interfaces move inward to the channel centre resulting in a larger radius of curvature and smaller surface tension. If the vapour pressure increases, the interfaces move outward to the corners leading to a smaller radius of curvature and higher surface tension along the interfaces. Thus, due to the triangular configuration, a restoring force develops to damp out pressure perturbations resulting in a stable annular flow mode. In contrast, this restoring force is absent in microchannels with rectangular cross section. The interfaces span the entire channel height, as shown in Figure 7.14-c, and the spanwise interface motion does not lead to changes in its curvature and surface tension. Thus, in a rectangular microchannel, the liquid film/vapour core interface fluctuates due pressure perturbations without a restoring force. Consequently, if the vapour pressure drops, the two interfaces bounding the vapour core can move inward without additional resistance. This movement can result in the merger of the two liquid films at some point along the channel (Figure 7.12-b), rendering the annular flow highly unstable. Of course, this instability is the result of not only shape but also size effect. It is straightforward to imagine a rectangular microchannel with a larger height as drawn in Figure 7.14-d. In such a case, known to develop in rectangular heat pipes, the liquid/vapour interface does not extend across the entire channel height; instead, the liquid is confined in each corner by an interface similar to the triangular configuration (Figure 7.14-b). Following the same restoring force argument, the annular flow in a deep rectangular channel should also be stable. It is interesting to note that the mechanism identified as responsible for the instability of annular flow in a shallow rectangular microchannel, is also the same physical mechanism responsible for: (i) the bubble departure demonstrated in Figure 7.6, and (ii) the establishment of intermittent flow downstream of the transition region shown in Figure 7.13.

7.4   Temperature field

Forced convection boiling is attractive since it ensures low device temperature for high power dissipation, or alternatively, it allows higher power dissipation for a given device temperature. Direct measurements of either the inner wall or the fluid bulk temperature distributions along a micro duct under forced convection boiling are not available yet due to the difficulty in integrating sensors at the desired locations. However, surface temperature measurements have been reported for a microchannel heat sink device with an array of 35 diamond-shaped microducts, (each about 40μm in hydraulic diameter) [82]. The spanwise temperature distributions were found to be uniform [77]; thus, only distributions along the device centreline are reported as shown in Figure 7.15. A significant reduction of the device temperature is evident in Figure 7.15-a, where the results for increasing water flow rate with constant heat flux of q=3.6W are summarized. Initially, the device temperature and its gradient, for a given power dissipation, is high since the heat is removed by free convection only. The maximum temperature of about 230°C is measured at the heater location. The device temperature drops sharply to about 115°C even with low water flow rate of 0.25mL/min (average velocity of 6.7cm/s within each duct). Increasing the water flow rate leads to further reduction of the device temperature to a level below the saturation temperature of about 100°C. This is expected since higher flow rate results in higher heat transfer rate. Consequently, the device internal energy, i.e. the device temperature, decreases.

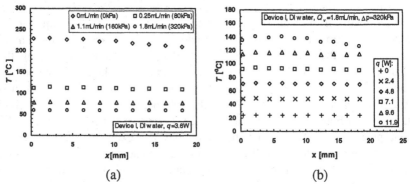

(a)   (b)

Figure 7.15: Streamwise temperature distributions along the diamond device centreline for different: (a) flow rate (q=3.6W), and (b) heat flux (Q_v=1.8mL/min).

Furthermore, the temperature distribution becomes more uniform with increasing flow rate, suggesting that the local heat transfer rate is highly non-uniform. It should be emphasized, though, that the flow is in single-liquid phase at the high flow rate and in two-phase at the low flow rate as indicated by the exit fluid void fraction. Hence, the heat-transfer mechanism changes character as the flow rate varies. The temperature distributions for increasing heat flux with constant water flow rate of Q_v=1.8mL/min (constant pressure drop of 320kPa) are summarized in Figure 7.15-b. Increasing the power input results in both higher device temperature and higher temperature gradient, as phase change takes place along the channels.

Surface temperature measurements have also been reported for microchannel heat sinks with 35 triangular microducts (each about 25μm in hydraulic diameter) [85]. The integrated heater in these devices, similar to the diamond-shaped devices (Figure 4.2-a), is located at one edge of the heat sink. Therefore, the temperature field is expected to depend on the flow direction. Temperature distributions along the device centreline for both flow directions are summarized in Figure 7.16. A typical set of data, for forward flow rate of Q_v=2mL/min, is shown in Figure 7.16-a. The distributions are fairly uniform for all input-power levels, except for the sensor nearest to the heater, although the vapour content of the exit fluid continuously varied. At the lower input power, q=5.1W, the flow was in liquid single phase throughout the microchannels. Increasing the power above q=11.4W resulted in two-phase flow within the channels. The fraction of vapour content increased with the input power until the critical heat flux (CHF) conditions developed at an input power level of q=29.9W. At this point the exit flow void fraction was unity as the entire channel flow was in single-phase vapour, with surface temperature slightly higher than 100°C, and no liquid droplets were detected by the microscope-camera system.

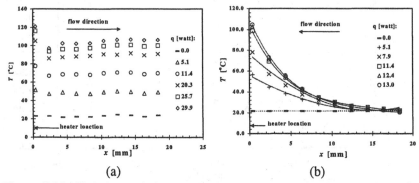

(a)   (b)

Figure 7.16: Variations of the streamwise temperature distributions along the triangular device centreline with increasing input power for: (a) forward flow (Q_v=2mL/min), and (b) backward flow direction (ΔP=240psi).

Reversing the flow direction, while keeping constant total pressure drop, resulted in dramatically different temperature distributions as evident in Figure 7.16-b. The temperature along the channel is highly non-uniform; it increases non-linearly from near room temperature at the channel inlet to its highest value at the channel outlet near the heater. This is in stark contrast to the fairly uniform temperature distributions for the forward flow. Although the driving pressure was kept constant, the flow rate increased with the input power, from 3mL/min for $q=11.4$W to 4.7mL/min for $q=13$W. The increased input power was consumed in vaporizing the additional liquid due to the higher flow rate. Hence, no temperature increase was recorded at this power range. The increased flow rate is due to: (i) decreasing liquid viscosity with increasing temperature, and (ii) liquid-to-vapour phase change as vapour viscosity is lower than liquid viscosity. In order to eliminate these effects, the system has to be modified such that the flow rate can be kept constant.

The measurements discussed thus far indicate that, for constant flow rate, increasing the heat flux always resulted in increasing surface temperature. Similar observations were reported for a microchannel heat sink with 10 nearly rectangular microducts (each 14μm in depth and about 120μm in width) for a relatively high flow rate [108]. However, at a lower water flow rate (about 0.0315mL/min), the surface temperature appeared to be constant over a small range of input power as demonstrated in Figure 7.17. The distribution is quite uniform for low input power with single-phase liquid flow. As the power increases, the temperature and its spatial gradient also increases. For higher input power, ranging from 1.83 to 2.04W, the local surface temperature is constant due to the liquid-vapour phase change.

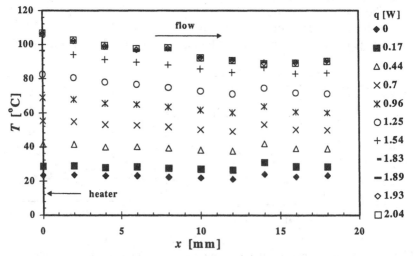

Figure 7.17: Variations of streamwise temperature distributions along the nearly rectangular device centreline with heat flux under flow rate of $Q_v=0.0315$mL/min.

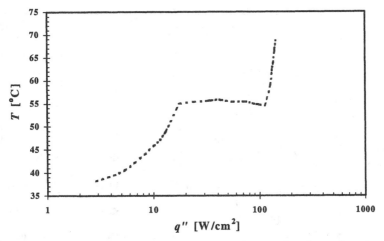

Figure 7.18: Exit fluid temperature as a function of the heat flux for a microchannel heat sink with circular channels, 510μm in diameter, under constant R-113 flow rate of 19mL/min and inlet subcooling of 20°C.

7.5   Boiling curves

The performance of a thermal system, such as a microchannel heat sink or a micro heat pipe, is perhaps best characterized quantitatively by its boiling curve of the system temperature as a function of the heat dissipation. Bowers and Mudawar [20] tested a microchannel heat sink with circular channels 510μm in inner diameter. They measured the exit fluid temperature as a function of the heat flux at a 19mL/min constant flow rate of R-113 dielectric coolant. The measurements for inlet subcooling of 20°C are plotted in Figure 7.18. The classical boiling plateau associated with latent heat of liquid-to-vapour phase change is evident as the exit temperature remains constant around 55°C for a wide range of heat flux, 20-110W/cm^2. Jiang *et al.* [82] measured the 2-D surface temperature field for a variety of heat sinks with diamond-shaped microchannels, either 40 or 80μm in hydraulic diameter. The power supplied to each device through a local heater was increased gradually, while the total pressure drop was held constant. The experiments were repeated for a variety of test parameters such as fluid driving pressure, size and number of channels of the heat sink. The flow rate and temperature distributions were registered after the system had reached a steady state under water inlet subcooling at room temperature (20°C). The spanwise temperature distributions were found to be uniform, and the streamwise temperature gradients were very small. Hence, only the average surface temperature as a function of the input power is reported for each case, and the results are summarized in Figure 7.19.

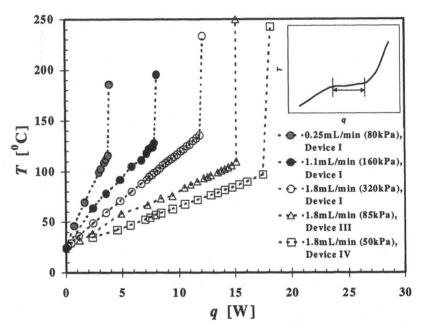

Figure 7.19: Boiling curves of average surface temperature as a function of the power dissipated by the heater in the diamond-shaped microchannel heat sinks for a variety of test parameters. (Inset: typical plateau in a standard boiling curve)

In all cases, the device average surface temperature increases monotonically, almost linearly, with the input power. At a certain power level, corresponding to Critical Heat Flux (CHF), the temperature increases sharply. The exit flow changes from single liquid phase (void fraction zero), through two-phase flow of liquid-vapour mixture, to a single vapour phase under CHF conditions (void fraction one). These boiling curves are in contrast to previously reported data plotted in Figure 7.18 for a larger microchannel 510μm in diameter. The typical boiling plateau, illustrated at the inset of Figure 7.19, has not been observed under all tested conditions. The plateau in the classical boiling curve is due to the saturated nucleate boiling, where bubbles continuously form, grow and depart such that the temperature is kept uniform and constant throughout the bulk fluid although the heat dissipation is increasing until CHF condition develops. Assuming the surface temperature faithfully represent the fluid temperature, the boiling curves in Figure 7.19 suggest that the saturated nucleate boiling may not develop in such micro ducts due to size effect. In order to understand the relationship between the flow pattern and temperature field, the boiling curve measurements should be correlated with visualizations of the corresponding flow patterns.

Similar boiling curve measurements were conducted for the triangular microchannel heat sinks with a hydraulic diameter of either 25 or 50μm [85]. Capping the microchannels with a glass substrate allowed in-situ monitoring of the flow pattern. The average temperature along the device centreline was again used to construct the boiling curves shown in Figure 7.20 for four different test conditions. All the results collapse together into a single curve if the surface temperature and the input power are normalized by their respective values at CHF conditions, i.e. T_{CHF} and q_{CHF}, and the reference room temperature, T_0. Although the curves are not linear, the temperature increases monotonically with the input power. No boiling plateau, where $\partial T/\partial q \sim 0$, can be observed similar to the curves shown in Figure 7.19. However, the slope gradually decreases toward zero as the flow approaches the CHF conditions, at which the device temperature increases sharply. Three flow regimes, marked in Figure 7.19, have been identified based on the phase-change modes detected by flow visualizations. For low input power level, $q/q_{CHF} < 0.4$, Region I, the flow in and out of the device was in liquid phase corresponding to the linear increase of the temperature with the input power. This indicates that almost all the increased power dissipation is transferred into the liquid increasing its internal energy, i.e. its temperature.

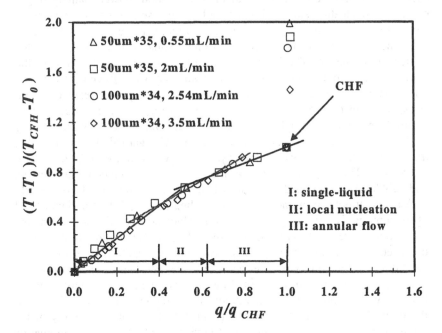

Figure 7.20: Normalized boiling curves of surface average temperature as a function of the power dissipation in the triangular-shaped microchannel heat sinks for different test conditions.

The first two-phase mode, local nucleation, was detected at an input-power as low as $q/q_{CHF}=0.4$. Bubbles could be seen forming along the channel walls at a few active nucleation sites. Bubble formation, growth and departure at a regular frequency have been documented (Figure 7.4). It is difficult to ascertain whether these are gas bubbles originated from dissolved gas in liquid (the solubility of air in water is about 1.5×10^{-5} at 20°C [111]), or vapour bubbles formed due to phase-change of the working liquid. On one hand, bubble activity was observed at device surface temperature of about 80°C, below the saturation temperature of water, indicating the possibility that the bubble source was dissolved air in water. On the other hand, the channel wall temperature was not measured; hence, it is not known whether locally the wall superheat requirement for nucleate boiling had perhaps been met. Since most of the nucleation sites became inactive after one or two runs of repeated experiments, the bubbles probably generated from residues of the fabrication process. Consequently, this local nucleation had negligible effect on the boiling curve, which stayed linear in the low input power range.

At moderate input-power levels, $0.4<q/q_{CHF}<0.6$, large bubbles were generated at the inlet/outlet common passages that connected the channel array to the device common inlet/outlet. The boiling activity at these larger passages became more intense with increasing input power, namely larger number of nucleation sites and higher bubble release frequency. As a result, the boiling-curve slope in Region II is smaller than the slope in Region I since the bubble-formation process consumed some of the input power due to the phase change. The bubbles typically grew to a size larger than the microchannel dimensions, and the upstream ones were forced through the microchannels resulting in a slug-flow mode. At a power level of about $q/q_{CHF}=0.6$, an flow mode was observed of a vapour core bounded by thin liquid films. Liquid droplets appeared in the vapour core with a wavy vapour-liquid interface. This mode should be regarded as an unstable, transition stage, since it was not always detected and short-lived whenever it did appear. The nucleation sites on the walls were completely suppressed, and bubble formation inside the microchannels could no longer be observed.

A stable annular flow mode was established (Figure 7.9) at a higher level of the input power, $0.6<q/q_{CHF}<0.9$. An interface between the liquid film and the vapour core was clearly distinguishable, and no liquid droplets were observed in the vapour core. At this power range, Region III, evaporation at the liquid-film/vapour-core interface dominated the heat transfer from the channel wall to the fluid. Additional input power is converted into: (i) latent heat of evaporation at the liquid-vapour interface due to the phase change, as the liquid films became thinner and thinner, and (ii) internal energy of the liquid film manifested by the increased device temperature. The combination of the two heat-transfer mechanisms resulted in monotonic temperature increase with decreasing slope as the input power increased.

As the input power approached the critical heat flux level, $q/q_{CHF} \to 1$, the liquid film started to dry out at certain portions on the wall, primarily at the channel exit region. Once the dryout process had been triggered, it proceeded rapidly such that the entire liquid in the channels was completely vaporized. This was accompanied by a dramatic rise in the device temperature, often melting the soldering material of the wirebonds and leading to a device failure.

Boiling curves were also reported for the micro heat sink with 10 nearly rectangular microchannels of about 24μm in hydraulic diameter [107]. The surface temperature at a few locations along the device centreline is plotted as a function of the power dissipation in Figure 7.21. Since the heater is located at one edge of the heat sink, the temperature decreases slightly with increasing distance from the heater. Nevertheless, all curves exhibit similar trend independent of the downstream location. Initially, the temperature at each location increases almost linearly with the input power as already reported. However, at power input close to the CHF level, the boiling curve characteristics depend on the water flow rate. It is interesting to note that for the low flow rate, Figure 7.21-a, the curves do exhibit the boiling-plateau feature, $\partial T/\partial q \cong 0$, associated with latent heat of liquid-to-vapour phase change. At the low input power range, the heat is removed from the system through both forced convection by the working fluid and natural convection from the device surface. When the temperature is high enough, a transition zone is established (Figure 7.13) separating between the upstream single-phase vapour and the downstream two-phase intermittent flow. Additional power is used to vaporize more liquid, and the vapour content increases as the transition region moves downstream with no temperature increase. When the transition region is at the exit, all the working fluid is in vapour phase giving rise to the onset of CHF, which is accompanied by a sharp increase of the temperature. However, for the high-flow rate, Figure 7.21-b, the curves are very similar to those shown in Figure 7.20 with no boiling plateau.

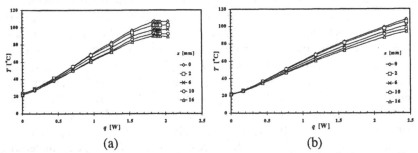

Figure 7.21: Boiling curves of local surface temperature along the device centreline in the nearly-rectangular microchannel heat sink for: (a) low (Q_v=0.0315mL/min), and (b) high flow rate (Q_v=0.063mL/min).

It is important to keep in mind that for proper boiling curves, either the fluid or wall temperature is required as a function of the heat removed by the working fluid. It is still difficult to integrate microsensors within the bulk fluid flow or on the inner surface of a microchannel to provide such temperature measurements. Therefore, only the device surface temperature measurements have been reported thus far. Moreover, a portion of the input power is stored as internal energy of the entire chip (and package), which may not be negligible. Still further, in cases of high device temperature and small flow rate, natural heat convection from the device surface to the ambient may not be negligible compared with the forced heat convection by the working fluid flow. Natural heat convection can be minimized by conducting the experiments in high vacuum environment and using liquids with low saturation temperature as the working fluids, while the internal energy within the solid device can be estimated. Moreover, with continuous development of micromachining technology, it may be possible to fabricate sensors at desired locations within microchannels and, thus, obtain more accurate boiling curves.

7.6   Critical heat flux

Critical heat flux (CHF) refers to the heat transfer limit due to a drastic reduction of the heat transfer coefficient and a sharp rise of the surface temperature, possibly leading to a failure of the device in which evaporation or boiling is occurring [134]. The high heat dissipation rates ($1MW/m^2$) required in the electronic industry, such as for power devices, can be achieved without exceeding CHF by using relatively low flow rate of boiling dielectric coolants in small diameter channels. On the other hand, cooling of devices, such as high-power lasers, demands ultra-high heat flux cooling technologies. The heat flux requirements of these devices are typically over $25MW/m^2$, far greater than those of electronic cooling. Since these heat fluxes exceed the CHF attainable with common flow boiling system, the ability to significantly enhance and accurately predict the magnitude of CHF is of paramount importance to ultra-high heat flux applications. High mass flow rate of subcooled flow boiling is required to maintain, under these extreme operating conditions, high heat fluxes in the nucleate boiling regime safely below CHF. Low cost, availability and high latent heat of vaporization render water the ideal coolant for these applications. A study of ultra-high CHF for subcooled flow boiling in small diameter tubes ascertained the parametric trends of CHF with respect to all important flow and geometrical variables. CHF increased with increasing mass flow rate, increasing subcooling, decreasing tube diameter, and decreasing heated length-to-diameter ratio. CHF was accompanied by physical 'burnout' of the tube wall near the exit, and tube material had little effect on the CHF magnitude.

Bowers and Mudawar [20] reported a distinct separation between mini- and micro-channel curves, which was explained as a result of the large difference in L/D ratio, L and D are the channel length and diameter, 3.94 for the mini-channel and 19.6 for the micro-channel. Consequently, they proposed the following CHF correlation:

$$\frac{q_c''}{Gh_{fg}} = 0.16We^{-0.19}\left(\frac{L}{D}\right)^{-0.54} \tag{7.8}$$

where q_c'' is the CHF based upon the heated channel inside area, G is the mass velocity and h_{fg} is the latent heat of evaporation. $We=G^2L/\sigma\rho$ is the Weber number, while σ and ρ are the liquid surface tension and density, respectively. They argued that the small diameter of the channels resulted in an increased frequency and effectiveness of droplet impact on the channel wall. This could have increased the heat transfer coefficient and enhanced the CHF compared to droplet flow regions in larger tubes. The small overall size of the heat sinks seemed to contribute to delaying CHF by conducting heat away from the downstream region undergoing partial or total dryout to the boiling region of the channel. Thus, higher heat flux was required for the onset of CHF conditions along the entire microchannel rather than just near the exit, and CHF was triggered even with negligible net vapour production.

Limited data is available for CHF in microchannels with hydraulic diameter smaller than 100μm. The exit flow void fraction under CHF condition is one, as the entire liquid passing through the heat sink changes phase into vapour. Therefore, it is reasonable that the critical heat flux increases linearly with the flow rate if most of the heat flux is converted into latent heat at about the saturation temperature. Indeed, experimental results for several microchannel heat sinks, plotted in Figure 7.22-a, roughly collapse along a linear curve of the input power at CHF, q_{CHF}, as a function of the water flow rate, Q_v, consistent with Equation 7.8 [82].

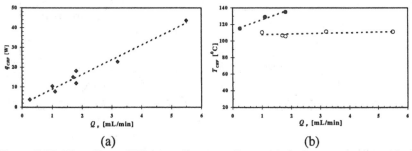

(a)   (b)

Figure 7.22: The effect of DI water flow rate Q_v on: (a) the input power at critical heat flux q_{CHF}, and (b) the corresponding average surface temperature T_{CHF}.

An important parameter associated with the CHF condition is the corresponding device temperature. The dependence of the average surface temperature under CHF condition, T_{CHF}, on the flow rate is shown in Figure 7.22-b for different heat sinks. For the heat sinks with larger channels, $D_h=80\mu m$, the CHF temperature depends neither on the flow rate nor on the number of channels. Furthermore, T_{CHF} is slightly higher than the saturation temperature of water under atmospheric pressure, 100°C. The higher CHF temperature may be due to the higher pressure, larger than 1atm, throughout the micro ducts. However, for the heat sink with smaller channels, $D_h=40\mu m$, the CHF temperature increases almost linearly with the water flow rate (full symbols in Figure 7.22-b), which cannot be attributed to the higher pressure. The difference between the two heat sinks is puzzling, and more experiments with wider flow-rate range are required.

7.7   Flow-regime map

There is a widely felt need to combine flow visualizations with boiling-curve measurements in order to devise a simple method for correlating the particular flow pattern likely to occur under a given set of forced convection parameters. One method to depict the various transitions is in the form of flow patterns maps, although it is impossible to represent the influence of all variables using only a two-dimensional plot. Indeed, a variety of maps have been proposed for macrosystems. For example, a map obtained from observations of low-pressure air-water and high-pressure stream-water flows in vertical small tubes is shown in Figure 7.23-a [60]. Another map for two-phase flows in horizontal and inclined tubes, shown in Figure 7.23-b, was developed based on various transition models [195].

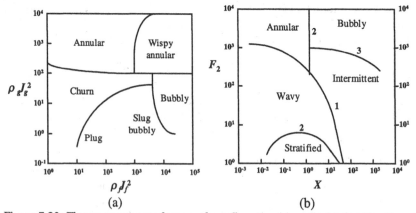

Figure 7.23: Flow pattern maps for two-phase flows in: (a) vertical (after Hewitt and Roberts, 1969), and (b) horizontal and inclined tubes (after Taitel and Dukler, 1976).

The superficial momentum fluxes of the vapour, $\rho_g J_g^2$, and of the liquid, $\rho_f J_f^2$, are plotted on the ordinates of the vertical flow pattern map. These fluxes have dimensions of pressure, and are formed with the vapour and liquid velocities, which would occur, if each of the two phases filled up the entire cross-section; they can be expressed in terms of the mass flow rate, Q_m, and the vapour quality, ξ, as: $J_g=\xi Q_m/\rho_g$ and $J_f=(1-\xi)Q_m/\rho_f$. It is clear that the momentum fluxes alone are not adequate to represent the influence of fluid physical properties or channel geometry. For the horizontal and inclined tubes, with inner diameter D and inclination angle θ, the boundaries between the various regimes in the flow pattern map are determined by the following non-dimensional expressions:

$$F_1 = \frac{J_g}{(gD\cos\theta)^{1/2}} \left(\frac{\rho_g}{\rho_f - \rho_g}\right)^{1/2} \qquad (7.9\text{-}a)$$

$$F_2 = \frac{\rho_g J_g^2 J_f}{g v_f \left(\rho_f - \rho_g\right)\cos\theta} \qquad (7.9\text{-}b)$$

$$F_3 = \left[\frac{(dP/dz)_f}{g\left(\rho_f - \rho_g\right)\cos\theta}\right]^{1/2} \qquad (7.9\text{-}c)$$

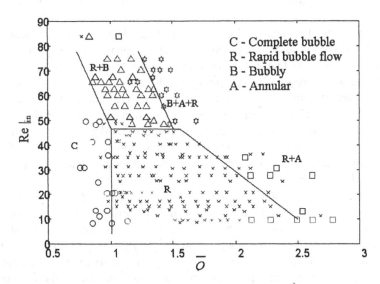

Figure 7.24: Flow regime map in triangular microchannels (courtesy of Prof. G. Hetzroni, 2002).

as a function of the Lockhart-Martinelli parameter X given by:

$$X^2 = \frac{(dP/dz)_f}{(dP/dz)_g} \qquad\qquad (7.10)$$

with the friction pressure drop $(dP/dz)_f$ of the liquid, if this alone were flowing through the tube, and the friction pressure drop $(dP/dz)_g$, if the vapour alone were flowing through the tube. Again, the heat flux, which is the driving force responsible all the flow patterns, does not enter the map explicitly. Indeed, the flow pattern maps must be regarded as no more than a rough guide, since the forces or mechanisms dominant in the different flow regimes are not precisely known. Moreover, the transitions among the various flow patterns are not sharp; clear boundaries with abrupt transitions do not exist. Hetzroni plotted a two-phase flow regime map for micro-channels, shown in Figure 7.24, based on experimental observations of convective boiling in triangular microchannels [58]. The ordinates are the inlet Reynolds number, Re_{in}, and the normalized heat flux, q. These are clearly the most dominant parameters in forced convection boiling and, thus, more appropriate as the map variables. This is an initial attempt to develop a flow regime map for microchannels, and more experimental work is needed to investigate the influence of other variables on the flow patterns.

7.8   Analytical modelling of forced convection boiling

Evaporating two-phase flow in a heated microchannel presents a formidable challenge for physical and mathematical modelling. It is not surprising, therefore, that to date very little analytical work has been reported on this subject. One of the first attempts to theoretically analyse a uniformly heated microchannel is the derivation of Peles *et al.* [146], based on fundamental principles rather than empirical correlations. The idealized flow pattern is sketched in Figure 7.25, where two distinct domains of liquid and vapour are separated by an infinitely thin evaporating front. The front propagates relatively to the fluid flow with the velocity u_F' equal to the linear rate of liquid evaporation. The front velocity in a frame of reference associated with the channel walls, u_F, equals: $|u_f|-|u_F'|$, as depicted in Figure 7.25, where u_f is the liquid velocity in an inertial coordinate system. Depending on the relationship between u_F' and u_f, the evaporation front can move either downstream $|u_f| > |u_F'|$, or upstream $|u_f| < |u_F'|$. In both cases, the flow in the microchannel is unsteady. At certain combination of the control parameters, such as inlet liquid velocity u_f and wall heat flux q, the flow may be statically steady in correspondence to the kinematic condition $|u_f| = |u_F'|$.

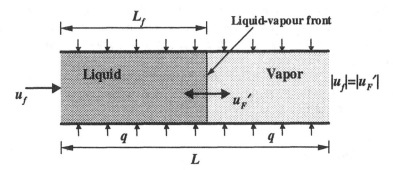

Figure 7.25: The physical model of convective boiling in a heated microduct with an evaporation front separating between the single-phase liquid and vapour flows.

In this model, the microchannel entrance flow is in single-phase liquid and the exit flow in single-phase vapour. The two phases are separated by the evaporation front at a location determined by the boundary conditions. The velocity, temperature and pressure distributions corresponding to such a flow are depicted in Figure 7.26. The velocity of the vapour increases monotonically along the channel, whereas the liquid velocity is almost constant due to its low thermal expansion. At the front, there is a jump in the flow velocity. The temperature distribution has a maximum within the liquid phase near the front as a result of the heat transfer from the wall to the liquid and heat removal due to liquid evaporation at the front. The pressure drops monotonically in both domains with a pressure jump at the front due to surface tension and phase change effect on the liquid-vapour interface.

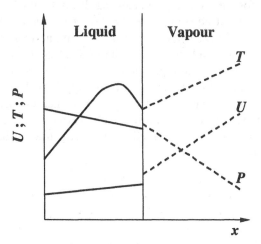

Figure 7.26: The velocity, temperature and pressure distributions along the axis of a heated microduct with an evaporation front (concave meniscus).

The mathematical model includes, on top of the standard equations of conservation and state for each phase, conditions corresponding to the interface surface. At steady flow in a heated microchannel the conditions at the evaporation front may be expressed by the continuity of mass, thermal fluxes on the interphase surface and the equilibrium of all acting forces. With reference to the evaporative meniscus the governing equations take the following form [146]:

$$\sum_{i=1}^{2} \rho_i u_i n_i = 0 \tag{7.11}$$

$$\sum_{i=1}^{2} \left(c_{p_i} \rho_i u_i T_i + k_i \frac{\partial T_i}{\partial x_i} \right) n_i = 0 \tag{7.12}$$

$$\sum_{i=1}^{2} \left(P_i + \rho_i u_i u_j \right) n_i = \left(\mu_{ij}^{(2)} - \mu_{ij}^{(1)} \right) n_j + \sigma \left(\frac{1}{r_1} - \frac{1}{r_2} \right) n_i + \frac{\partial \sigma}{\partial x_i} \tag{7.13}$$

where μ_{ij} is the tensor of viscous tension, r_1 and r_2 are the general curvature radii of the interphase surface, n_i and n_j correspond to the normal and tangent directions, while $i=1$ for the vapour and $i=2$ for the liquid phase. This model, though simplified a great deal, clearly demonstrates the complexity of the theoretical modelling, which will have to be developed in order to obtain meaningful results. Under the assumption of a quasi-one-dimensional flow, a set of equations for the average parameters has been derived in order to solve the resulting system of equations. Steady flow was found to be possible if $q_f > Ja$; where $q_f = q \cdot L/(\rho_f u_f c_{pf} T_f)_{x=0}$ is the wall heat flux normalized by the inlet conditions, while the Jakob number is the defined as $Ja = h_{fg}/(c_{pf} T_f)_{x=0}$. For steady flow, the extent of the liquid domain as well as the maximum liquid temperature was found to decrease with increasing heat flux. A comparison between analytical calculations and experimental measurements requires a more refined physical model, allowing for periodic oscillations of the evaporation front, and more detailed data of streamwise distributions of relevant flow properties.

Chapter 8

Unsteady Convective Heat Transfer in Micro Ducts

The classification between steady and unsteady convective boiling in internal flows refers only to the controlled parameters such as heat flux or mass flow rate. Single-phase heat convection flows under steady boundary conditions are expected to be steady, especially in microsystems as discussed in Chapter 6, in the sense that all flow properties are constant in time and may vary only in space. However, some properties of two-phase forced convection flow, under steady heating, may not be steady such as the oscillations of the liquid-vapour interface discussed in Chapter 7. Then, even if all boundary conditions are steady, some flow properties may vary in space and time due to flow instabilities that develop within the system. In this context, unsteady convective heat transfer to be discussed hereafter refers to controlled variations of certain parameters with respect to time. In particular, the response of the microsystem to a step or a periodic change of one of the input parameters is of interest as in control theory. Experimental and analytical studies of unsteady heat convection in microducts have just begun and some preliminary results are discussed [84,176].

8.1   Transient response to a step-current input

The most convenient parameter to vary in time, in theoretical analyses and numerical simulations, is the heat flux boundary condition. Similarly, when a heater is used in experimental work, the time-dependent electric current is the easiest parameter to utilize for controlling the heat flux input. The transient temperature field, due to a step-current input, was analysed for a diamond microchannel heat sink, described in Chapter 4, under two test conditions: (i) no flow within the microchannels (natural convection), and (ii) water flow through the microchannels (forced convection) [79].

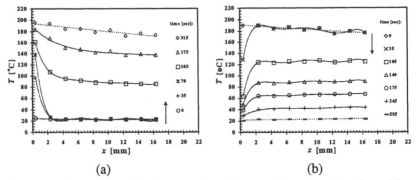

(a)   (b)

Figure 8.1: Temperature distributions along the centreline of a microchannel heat sink due to a step-current input at a few time steps under natural convection (no water flow) during the (a) heating-up, and (b) cooling-down process (q=3.2W).

8.1.1   Natural convection

The time response of the packaged device with no flow, i.e. under natural convection from the chip to the ambient air, was investigated. The heater was powered on by a step current signal. The resulting step input power, q, ranged from 0.2W to 3.2W. After the device had reached a steady state, for each step level, the power was suddenly turned off. The instantaneous temperature distributions along the device centreline at a few time steps are plotted in Figure 8.1. In the heating-up process, Figure 8.1-a, the chip temperature near the heater increases sharply within the initial few seconds while it stays constant further away. Then, the device surface temperature increases everywhere with a nonlinear spatial distribution, and a steady state of nearly linear temperature distribution is obtained after more than 5min. In the cooling-down process, Figure 8.1-b, the temperature at the heater drops quickly within half a minute while the temperature elsewhere barely changes. After almost 2min, the temperature of the entire device starts decreasing, and it reaches the steady state of uniform, ambient temperature after almost 10min. During the entire cooling-down process, the temperature distribution along the device is almost uniform except at the heater location.

The time evolution of the local transient temperature due to a step current input is summarized in Figures 8.2-a and b for the heating up and cooling-down process, respectively. All the temperature profiles, for different locations and different power levels, collapse onto a single exponential curve when a local time constant, either τ_h $(x;q)$ or $\tau_c(x;q)$ and the local steady-state temperature, $T_{ss}(x;q)$, are used as the normalizing parameters. Evidently, under natural convection, the device temperature response to a step power input is either exponential rise for heating-up process or fall for cooling-down process. This is typical to the response of a first order system.

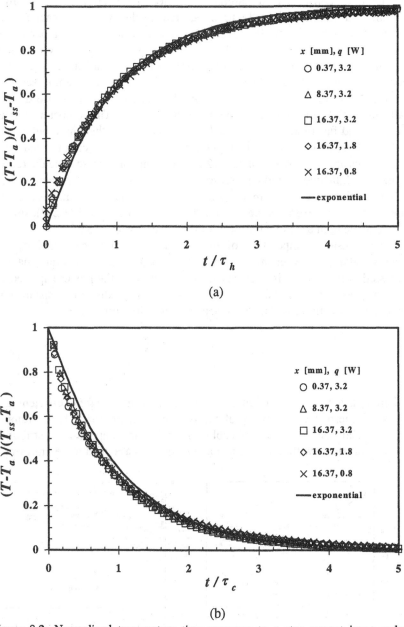

Figure 8.2: Normalized temperature time response to a step-current input under natural convection for the (a) heating-up, and (b) cooling-down process at a few locations along the device for various input power levels.

The time constant is a natural parameter characterizing a first-order system, and it is defined as the time required for the system to reach 63.2% of its steady-state level. However, the time constant is not a parameter applicable to characterize higher-order systems. A more convenient parameter is the temperature rise-time for heating-up or fall-time for cooling down process, which is defined as the time required for a system to settle to within 10% of the steady-state level. For a first-order system, the rise-time or fall-time is 2.3 times its corresponding time constant. The temperature rise-time, t_h, and fall-time, t_c, are plotted as a function of either the input power level or the sensor location in Figures 8.3-a and b, respectively, and both time scales are in the range of 150-220s. The temperature rise-time decreases while the fall-time slightly increases with increasing input power, Figure 8.3-a. Furthermore, the temperature rise-time and fall-time increase along the device (away from the heater) for the higher power level (3.2W), Figure 8.3-b, while both are quite uniform for the lower power (0.8W).

The transient temperature of the thermal system is governed by the energy balance between heat conduction through the Si substrate, natural convection from the chip surfaces to the ambient and the power supplied by the integrated heater. The ratio of the internal conduction resistance to external convection resistance is defined as the Biot number:

$$Bi = \frac{hL_s}{k_s} \tag{8.1}$$

where L_s is a characteristic length scale, h is the heat transfer coefficient, and k_s is the thermal conductivity of the Si substrate. For $Bi<0.1$, the lumped system analysis model is applicable, in which the temperature distribution during transients within the solid at any instance is assumed to be uniform.

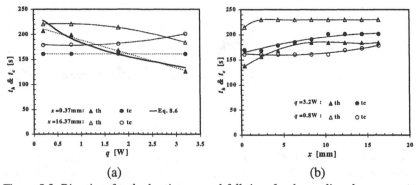

Figure 8.3: Rise time for the heating-up and fall-time for the cooling-down process as a function of: (a) input power level, and (b) sensor distance from the heater.

The measured time response suggests that the transient behaviour is dominated, for a given power dissipation level, by the balance between the heat stored as thermal energy within the device and the heat transferred from the device surface to the ambient by natural convection. Several empirical correlations have been suggested for the coefficient of heat transfer under natural convection, h_N, such as the following expression [71]:

$$h_N = 1.32 \left(\frac{T - T_a}{L_s} \right)^{1/4}$$
(8.2)

where T is the device surface temperature and T_a the ambient temperature. In order to account for the effect of the package, to which the device chip is attached, the size of the package, $38 \times 32 \times 1.5 \text{mm}^3$, is used in the estimation rather than the chip size, $25 \times 12 \times 1 \text{mm}^3$. The value of the heat transfer coefficient in the range of 6-12 $\text{W/m}^2\text{K}$ depends on the temperature, which is a function of the power input. Thus the higher the power input, the larger the heat transfer coefficient. The corresponding Biot number is in the order of 10^{-3}. Hence, the lumped system analysis model is valid, and the device temperature gradients within the substrate can be neglected. Consequently, the governing equation can be written as:

$$\rho_s c_{p_s} V_s \frac{dT}{dt} + h_N A_N \left(T - T_a \right) = I^2 R$$
(8.3)

where T is the device temperature, which is a function of time t only. ρ_s and c_{ps} are the density and specific heat of the Si substrate, respectively, V_s is the volume of the chip, A_N the surface areas for the natural convection heat transfer, and $I^2 R$ is the power supplied by the heater.

Equation (8.3) is a first-order ordinary differential equation, and can be easily solved subject to the proper initial condition for either the heating-up or the cooling-down process:

$$T\big|_{t=0} = T_a \qquad \text{(heating-up)}$$
(8.4-a)

$$T\big|_{t=0} = T_{ss} \qquad \text{(cooling-down)}$$
(8.4-b)

The corresponding solutions are given by:

$$\frac{T-T_a}{T_{ss}-T_a}=1-\exp\left(-\frac{2h_N}{\rho_s c_{p_s}\delta}t\right) \qquad \text{(heating-up)} \qquad (8.5\text{-a})$$

$$\frac{T-T_a}{T_{ss}-T_a}=\exp\left(-\frac{2h_N}{\rho_s c_{p_s}\delta}t\right) \qquad \text{(cooling-down)} \qquad (8.5\text{-a})$$

where δ is the Si substrate thickness, $\delta=2V_s/A_N$, and T_{ss} is the steady-state temperature. Thus, the time constant associated with this model for either heating-up or cooling-down process is given by:

$$\tau_h=\tau_c=\frac{\rho_s c_{p_s}\delta}{2h_N} \qquad (8.6)$$

The estimated time scales, based on Equation (8.6), depend on the heat transfer coefficient, which in turn is a function of the input power. Since the heat transfer coefficient increases with increasing device temperature, the rise-time and the fall-time should decrease with increasing input power. The estimated decrease of the rise-time with the increased power input, plotted in Figure 8.3-a, agrees with the measured trend for the heating-up process. However, the measured fall-time for the cooling-down process is essentially independent of the power input, indicating that the simplified theoretical model is not adequate for the detailed analysis of the real system.

Equation 8.3 also predicts that the steady-state temperature should be a linear function of the input power as follows:

$$T_{ss}-T_a=\frac{I^2R}{2(W+\delta)L_d h_N}\propto q \qquad (8.7)$$

where W and L_d are the width and length of the device, respectively. Indeed, the measured device steady-state temperature, plotted in Figure 8.4-a, increases linearly with the input power with a slope of 52.5K/W at all locations. The estimated steady-state temperature from Equation 8.7, with the slope of 55.5K/W, fits the experimental data reasonably well.

The steady-state temperature distribution along the device centreline x, though neglected in the time-dependent calculations, can still be determined. It is essentially a result of the balance between conduction through the silicon substrate, natural convection from the device surface to the ambient air and the heat input to the system as follows:

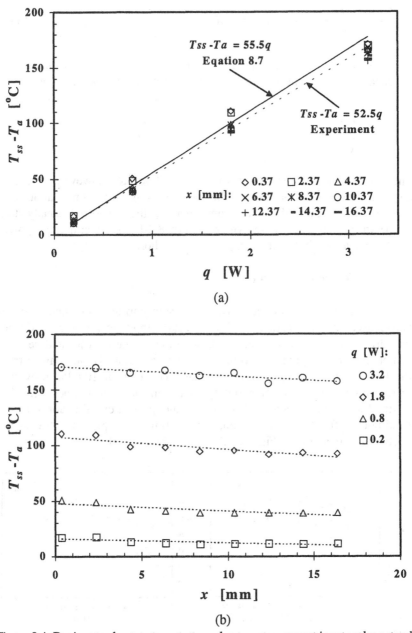

Figure 8.4: Device steady-state temperature, due to a step-current input under natural convection, as a function of: (a) the input power at a few locations along the heat sink, and (b) the distance from the heater for different input power level.

$$k_s \delta \frac{\partial^2 T}{\partial x^2} - 2h_N \left(T - T_a \right) = I^2 R \qquad (8.8)$$

The temperature is then expected to decrease exponentially as a function of the distance from the heater with a characteristic length scale, L_c, given by:

$$L_c = \left(\frac{2h_N}{k_s \delta} \right)^{-1/2} \qquad (8.9)$$

The length scale is about 50mm for the tested device. However, the actual distance between the heater and the last sensor is only 18mm, about one third of the estimated scale. Consequently, the measured steady-state temperature distributions, shown in Figure 8.4-b for several input-power levels, appear nearly linear since the device is short.

8.1.2   Forced convection

The time response of the device to a step current input under forced convection was also characterized for the same input power range. DI water was driven into the microchannels through the device inlet located near the heater, shown in Figure 4.2, while the outlet was about 18.6mm away from the heater. The driving pressure was kept constant, at 100kPa, providing water volume flow rate of about Q_v=4mL/min at room temperature. The device transient temperature distributions along the flow direction for both heating-up, after heater switch-on, and cooling-down process, after heater switch-off, are shown in Figures 8.5-a and b, respectively.

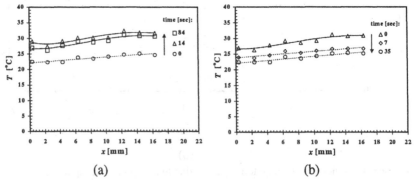

(a)   (b)

Figure 8.5 Temperature distributions along the centreline of a microchannel heat sink due to a step-current input at a few time steps under forced convection during the (a) heating-up, and (b) cooling-down process (q=1.8W, Q_v=4mL/min).

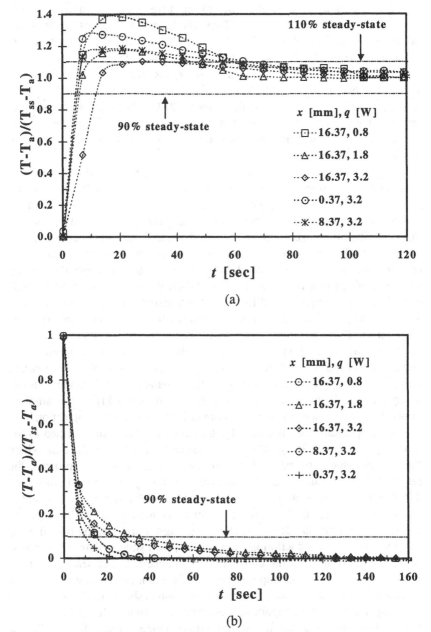

Figure 8.6: Normalized temperature time response to a step-current input under forced convection for the (a) heating-up, and (b) cooling-down process at a few locations along the device for various input power levels.

In the heating-up process, the temperature increases to its maximum level in a very short time of (about 14s). Then, the temperature drops gradually to the steady state (in about 84s). During the cooling-down process, the temperature drops to the steady-state temperature within 35s. Unlike the natural convection case, the temperature increases along the device, from the inlet to the outlet, mainly due to the forced convection cooling effect. The temperature of the DI water at the inlet is 20°C. Thus heat is transferred between the channel walls and the forced convection fluid due to the temperature difference, and the temperature of the cooling water increases with downstream distance.

The typical time response of device transient temperature to a step current input under forced convection is shown in Figure 8.6. In the heating-up process, Figure 8.6-a, the temperature demonstrates a clear overshoot before it reaches the steady-state temperature. The temperature increases rapidly from its initial to its maximum value, and then monotonically decreases to its eventual steady-state value. At the initial sharp temperature increase, the heat transfer mode is dominated by heat conduction through the device substrate, which is a faster mechanism. When the wall temperature is high enough, a large temperature difference exists across the wall-fluid interface. As a result, forced convection becomes the dominant heat transfer mechanism. Then, the fluid temperature increases and the wall temperature decreases till the system reaches its steady state. This is different from numerical simulations of the transient response where the temperature monotonically approaches the steady-state temperature [176]. In the cooling-down process, Figure 8.6-b, the temperature drops quickly to the ambient temperature due to the enhanced cooling effect by the forced convection.

The temperature rise-/fall-time for this microsystem is still defined as the time required for attaining 90% of the steady-state value. In the case where the system exhibits overshoot behaviour, e.g. in the heating-up process, the response time is usually stated as the time for the system to settle to within ±10% of the steady -state value. Of course, the rise-/fall-time will be equal to the response time, if the system response is monotonic such as in the cooling-down process. However, in the heating-up process, the rise-time, t_h, dominated by the conduction mechanism, is much smaller than the response time, t_{hres}, dominated by the convection mechanism. The dependence of the device rise-/fall-time and response time on the input power as well as on the distance from the heater under forced convection is shown in Figures 8.7 and 8.8, respectively. The temperature rise-time increases with the input power and is less than 10s for the tested power range, Figure 8.7, while the response time for heating-up process is in the range of 20-60s. On the other hand, the fall-time for the cooling-down process, equal to the response time, is less than 10s. Both the rise- and fall-time are almost uniform along the entire chip as shown in Figure 8.8. Slight decrease of the heating-up response time and slight increase of the fall-time are observed with increasing distance from the heater under forced convection conditions. In comparison with the natural convection case, the response time drops from 170 to 50s for the heating-up process, and from 190 to 20s for the cooling-down process. Moreover, the temperature rise-time under forced convection is less than 10s through the whole chip. Thus, the forced convection reduces the microsystem thermal time scale by one order of magnitude.

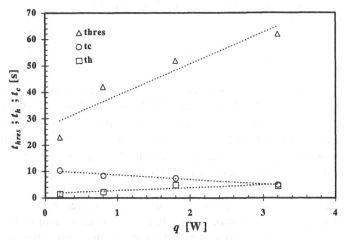

Figure 8.7: The rise and response time for the heating-up and the fall time for the cooling-down process as a function of the input-power step under forced convection.

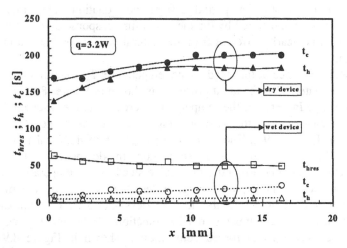

Figure 8.8: A comparison between the time scales under forced- (wet device) and natural-convection (dry device) for the same step-current input (q=3.2W).

The transient temperature of the thermal system is governed by the energy balance between the heat conduction through the Si substrate, natural convection from the chip surfaces to the ambient, forced convection by channel fluid flow and the heater power input. The device spanwise temperature distribution has been found to be uniform [77]. Thus, the one-dimensional, time-dependent, governing energy equation accounting for forced and natural convection heat transfer takes the form:

$$\rho_s c_{p_s} V_s \frac{\partial T}{\partial t} + h_N A_N (T - T_a) + h_F A_F (T - T_b) +$$

$$\rho_f c_{p_f} Q_v t \frac{\partial T_b}{\partial t} = I^2 R + k_s V_s \frac{\partial^2 T}{\partial x^2} \qquad (8.10)$$

where T_b is the fluid bulk temperature, ρ_f and c_{pf} are the density and specific heat of the working fluid, Q_v is the fluid volume flow rate through the microchannels, A_F and h_F are the surface area and the heat-transfer coefficient for the forced convection heat transfer, respectively. The heat transfer coefficient is estimated to be about 1000W/m^{2}K for liquid forced convection [148]. The corresponding Biot number is about 0.17, larger than the critical value of 0.1. Hence, the lumped system analysis model is not valid, and conduction cannot be neglected. Furthermore, both the substrate temperature and the fluid bulk temperature are time dependent and coupled. The resulting governing equation is fairly complicated, especially in light of the complex boundary condition, which is neither uniform heat flux nor constant temperature. The transient behavior, therefore, can not be described by a simple, first-order partial differential equation. The higher-order behaviour is demonstrated by the measured time response depicted in Figure 8.6, which is clearly different from the natural convection behaviour shown in Figure 8.2.

The dependence of the steady-state temperature, T_{ss}, on either the distance from the heater, x, or the input power, q, is shown in Figures 8.9-a and 8.9-b, respectively. In general, the temperature increases as the distance from the heater increases, and the positive slope increases with the input power, shown in Figure 8.9-a. The fluid absorbs heat from the local heater at the inlet, and transfers heat to the substrate along the downstream. Under forced convection with any flow rate, the device steady-state temperature level for the entire tested input power range is less than 35°C, and the maximum temperature difference over the device is less than 5°C. The comparison of device steady-state temperature as a function of the input power under natural convection and forced convection is shown in Figure 8.9-b. The temperature under forced convection also increases linearly with the input power, similar to the natural convection case. However, the slope for the device under forced convection, about 2.6K/W, is more than one order of magnitude lower than the slope for the device under natural convection, about 52.5K/W, due to the enhanced forced-convection heat transfer.

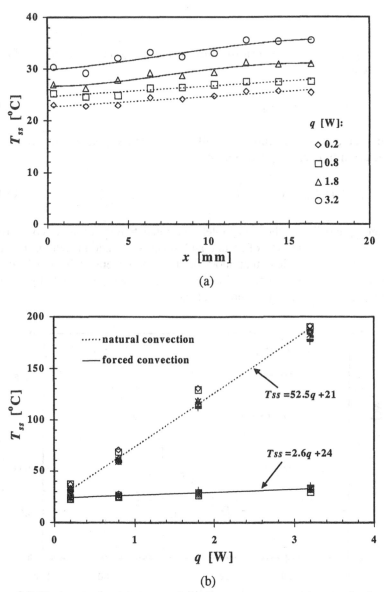

(a)

(b)

Figure 8.9: Device steady-state temperature, due to a step-current input under forced convection, as a function of: (a) the distance from the heater for different input power level, and (b) the input power at a few locations along the heat sink.

The relative forced, by the working fluid, and natural heat transfer rate, to the ambient, can be estimated. Assuming the device steady-state temperature is uniform, the energy balance of the thermal system can be written as:

$$I^2 R = q = q_N + q_F \qquad (8.11)$$

where q_N and q_F are the corresponding heat dissipation by natural and forced convection. Thus,

$$q_N = h_N A_N (T - T_a) \qquad (8.12)$$

The heat transferred from the wall to the working fluid must equal the increase in fluid enthalpy, therefore:

$$q_F = \rho_f Q_v c_{p_f} (T_o - T_i) \qquad (8.13)$$

where T_i and T_o are the water inlet and outlet temperatures. A comparison between the contribution of natural and forced convection is demonstrated in Figure 8.10. It is clear that most of the heat generated by the heater is removed by forced convection, while the heat removed by natural convection is only 2-5% of the total heat dissipation.

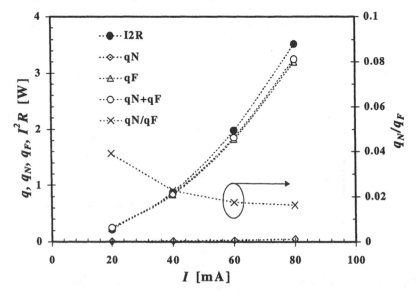

Figure 8.10: Relative contribution of the natural and forced convection heat transfer rate as a function of the step-current level under forced convection condition.

8.2   Unsteady response to periodic forcing

The transient response of the device temperature to a periodic input power rather than to a step function is of interest in many applications [84]. A sinusoidal voltage signal, $V(t)$, given by:

$$V(t) = \frac{V_{max} + V_{min}}{2} + \frac{V_{max} - V_{min}}{2} \sin(2\pi ft) \qquad (8.14)$$

is applied to the heater. Here, f is the driving signal frequency, while V_{max} and V_{min} are the voltage maximum and minimum levels, respectively. The periodic temperature response can be modelled as follows:

$$T(t) = \frac{T_{max} + T_{min}}{2} + \frac{T_{max} - T_{min}}{2} \sin(2\pi ft + \varphi) \qquad (8.15)$$

where φ is the phase shift. Both the temperature maximum, T_{max}, and minimum, T_{min}, depend on the parameters of the input voltage signal as well as on the forced convection flow rate and material properties of the microsystem (solid and fluid).

A typical device transient temperature response to a sinusoidal voltage input of varying frequency is shown in Figure 8.11. It is clear that the curves are not perfect sinusoidal functions, since the heating-up rise time is much shorter than the cooling-down fall time.

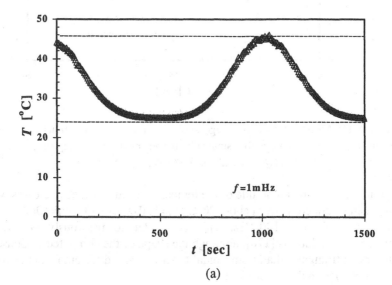

(a)

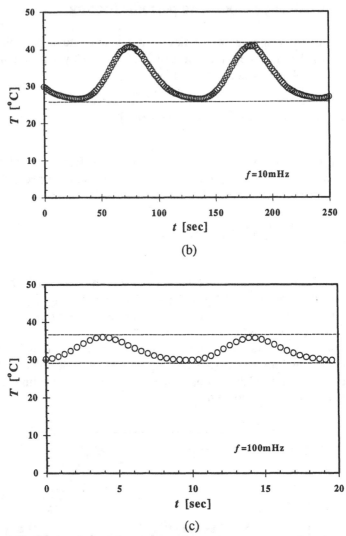

(b)

(c)

Figure 8.11: Measured spatial-average surface temperature as a function of time under forced convection due to a sinusoidal power input with a frequency of: (a) 1, (b) 10, and (c) 100mHz. (Q_v=0.4mL/min, V_{dc}=16V, V_{pp}=32V)

Furthermore, the amplitude of the temperature fluctuations decreases with increasing frequency, from about 20°C at f=1mHz, to 16°C at f=10mHz, and to about 7°C at f=100mHz. This is similar to previously performed numerical simulations [176], although the shape of the computed response is closer to a triangular than a sinusoidal function, very different from the wave form obtained in the experiment.

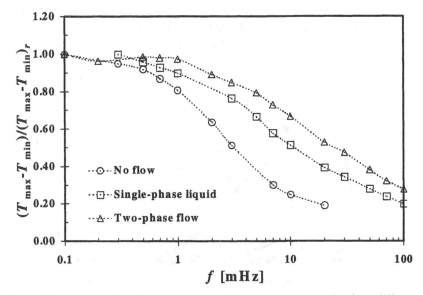

Figure 8.12: A comparison between measured frequency response for three different flow conditions: no flow, single-phase liquid, and liquid-vapour two-phase flow.

The frequency response of the device temperature for three cases: natural convection, single-phase liquid forced convection and liquid-vapor two-phase forced convection is summarized in Figure 8.12. The normalized peak-to-peak value of the temperature response $(T_{max}-T_{min})/(T_{max}-T_{min})_r$ decreases as a function of the input frequency; $(T_{max}-T_{min})_r$ being the peak-to-peak reference value of the temperature oscillations for the lowest tested frequency $f_r=0.1 \text{mHz}$. The curves demonstrate that the frequency response of the device under natural convection can be improved by forcing liquid through the microchannels, which enhances the heat transfer rate. Further improvement can be achieved by allowing fluid phase change inside the channels due to the enhanced heat transfer of evaporation.

The effect of flow rate on the transient temperature is shown in Figure 8.13-a for an input voltage signal with $f=10 \text{mHz}$. It is clear that regardless of the fluid phase conditions, i.e. whether the flow is in single-phase liquid, liquid-vapour, or single-phase vapour, all the curves are sinusoidal with a frequency of 10mHz. The dependence of the maximum and minimum temperature on the flow rate is shown in Figure 8.13-b. As long as the flow rate is high enough to keep the fluid in single-phase liquid or liquid-vapour mixture, the maximum temperature decreases gradually with increasing flow rate, while the minimum temperature is almost constant. However, both the minimum and the maximum temperature increase dramatically at low flow rate, when the exit fluid is in single-phase vapour. Incremental decrease in the flow rate leads to a single-phase vapour throughout the channels during the entire thermal cycle. This unstable condition results in a complete dryout of the device with very high temperature, damaging the device.

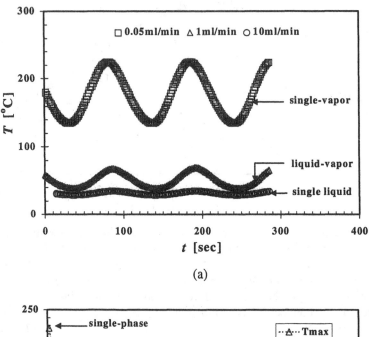

(a)

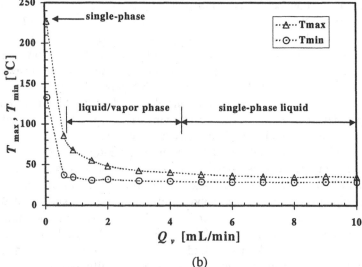

(b)

Figure 8.13: The effect of liquid flow rate on the: (a) instantaneous device temperature, and (b) maximum and minimum temperatures due to sinusoidal input power. (V_{dc}=40V, V_{pp}=32V)

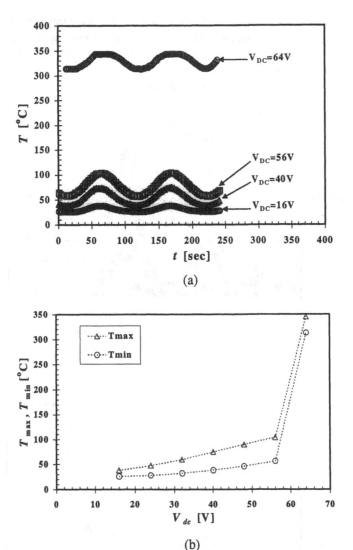

Figure 8.14: The effect of the dc input voltage on the: (a) instantaneous device temperature, and (b) maximum and minimum temperatures due to sinusoidal input power. (V_{pp}=32V, Q_v=0.7mL/min)

The input power level can be controlled by varying either the average, $V_{dc}=(V_{min}+V_{max})/2$, or the peak-to-peak value, $V_{pp}=V_{max}-V_{min}$, of the voltage signal applied to the heater. The effect of the dc level of the sinusoidal input power on the transient temperature response is presented in Figure 8.14-a. The device temperature increases with increasing dc power level, while the amplitude of the temperature oscillations does not change much for all fluid phase conditions. The dependence of the maximum or minimum temperature on V_{dc}, shown in Figure 8.14-b, is very similar. As soon as the flow is in single-phase vapour, around V_{dc}=64V, the temperature exceed 300°C, dryout occurs and the device can no longer function.

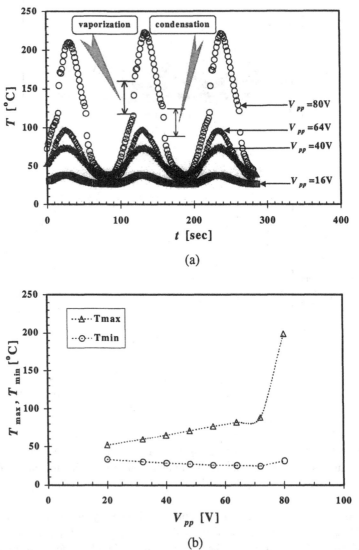

Figure 8.15: The effect of the input voltage amplitude on the: (a) instantaneous device temperature, and (b) maximum and minimum temperatures due to sinusoidal input power. (V_{dc}=40V, Q_v=0.8mL/min)

The effect of the peak-to-peak level of the sinusoidal input power, V_{pp}, on the transient temperature response is depicted in Figure 8.15-a. At the highest input amplitude, $V_{pp}=80V$, the temperature response is not sinusoidal anymore. A sudden increase during the heating-up and a sudden decrease during the cooling-down portion of the cycle can be observed at a device temperature of about 120°C. This result suggests that both a complete vaporization and a condensation process take place in a single thermal cycle, which prevent the occurrence of a dryout. Thus, temperature cycling may allow the device to reach high temperature while protecting it from the damage associated with dryout. Interestingly, the minimum temperature does not change as the power amplitude increases, as shown in Figure 8.15-b. In contrast, the maximum temperature increases almost linearly with the input amplitude. However, the maximum temperature increases dramatically when the vaporization-condensation conditions develop. A large peak-to-peak temperature range, up to 180°C, can be achieved in the amplitude-controlled mode. This operation mode provides a potential approach to carry out thermal cycling when a large temperature difference within a single cycle is required such as in chemical or biological applications.

8.3　Numerical simulations of transient response

The problem of unsteady forced convection heat transfer with phase change is much more complicated than the steady one due to the additional time-dependent terms that must be considered. Though a complete analytical solution is still lacking, attempts have been made to simplify the physical model so that, at least, the corresponding governing equations can be solved numerically. Rujano and Rahman [176] introduced such an approach de-coupling the momentum and energy equations so that a simultaneously developing steady fluid flow and transient heat transfer model is considered.

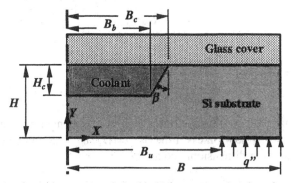

Figure 8.16: A schematic cross section of half the symmetric microchannel heat sink used for the numerical simulations with the heat source at the bottom right corner.

A schematic illustration of their physical model is shown in Figure 8.16. The fluid is assumed to be Newtonian and the flow is laminar and incompressible. The flow rate and heat flux are such that material properties remain approximately constant and buoyancy effects are negligible. The fluid enters the channel with a uniform temperature T_i and a uniform velocity W_i. A constant heat flux is prescribed at the bottom along the entire length L, as indicated in Figure 8.16. The part of the bottom surface not covered by the heat source, with input heat flux q'', is assumed insulated. The applicable non-dimensional differential equations for the conservation of mass, momentum, and energy in a rectangular coordinate system are [176]:

$$\frac{\partial u}{\partial x} + \frac{\partial v}{\partial y} + \frac{\partial w}{\partial z} = 0 \tag{8.16}$$

$$u\frac{\partial u}{\partial x} + v\frac{\partial u}{\partial y} + w\frac{\partial u}{\partial z} = -\frac{\partial p}{\partial x} + \frac{1}{Re}\left(\frac{\partial^2 u}{\partial x^2} + \frac{\partial^2 u}{\partial y^2} + \frac{\partial^2 u}{\partial z^2}\right) \tag{8.17}$$

$$u\frac{\partial v}{\partial x} + v\frac{\partial v}{\partial y} + w\frac{\partial v}{\partial z} = -\frac{\partial p}{\partial y} + \frac{1}{Re}\left(\frac{\partial^2 v}{\partial x^2} + \frac{\partial^2 v}{\partial y^2} + \frac{\partial^2 v}{\partial z^2}\right) \tag{8.18}$$

$$u\frac{\partial w}{\partial x} + v\frac{\partial w}{\partial y} + w\frac{\partial w}{\partial z} = -\frac{\partial p}{\partial z} + \frac{1}{Re}\left(\frac{\partial^2 w}{\partial x^2} + \frac{\partial^2 w}{\partial y^2} + \frac{\partial^2 w}{\partial z^2}\right) \tag{8.19}$$

$$\frac{\partial \theta_f}{\partial \tau} + u\frac{\partial \theta_f}{\partial x} + v\frac{\partial \theta_f}{\partial y} + w\frac{\partial \theta_f}{\partial z}$$

$$= \frac{1}{RePr}\left(\frac{\partial^2 \theta_f}{\partial x^2} + \frac{\partial^2 \theta_f}{\partial y^2} + \frac{\partial^2 \theta_f}{\partial z^2}\right) \tag{8.20}$$

$$\frac{\partial \theta_s}{\partial \tau} = \frac{\alpha_s/\alpha_f}{RePr}\left(\frac{\partial^2 \theta_s}{\partial x^2} + \frac{\partial^2 \theta_s}{\partial y^2} + \frac{\partial^2 \theta_s}{\partial z^2}\right) \tag{8.21}$$

x, y, and z are the coordinates normalized by the channel hydraulic diameter D_h, while the dimensionless time is $\tau = tW_i/D_h$. The temperature T and pressure P are non-dimensionlized as,

$$p = (P - P_o)/(\rho W_i^2/2) \quad ; \quad \theta = (T - T_i)/(q'' B/k_f) \tag{8.22}$$

The boundary conditions are the following,

At $z = 0$: $u = 0$, $v = 0$, $w = 1$, $\theta_f = 0$ (in the fluid region, inlet)

$$\frac{\partial \theta_s}{\partial z} = 0 \text{ (in the solid region)} \qquad (8.23\text{-}a)$$

At $z = L/D_h$: $p = 0$ (in the fluid region, outlet)

$$\frac{\partial \theta_s}{\partial z} = 0 \text{ (in the solid region)} \qquad (8.23\text{-}b)$$

At $x = 0$: $u = 0$, $\dfrac{\partial v}{\partial x} = 0$, $\dfrac{\partial w}{\partial x} = 0$, $\dfrac{\partial \theta_f}{\partial x} = 0$ (in the fluid region)

$$\frac{\partial \theta_s}{\partial x} = 0 \text{ (in the solid region)} \qquad (8.23\text{-}c)$$

At $x = B/D_h$: $\dfrac{\partial \theta_s}{\partial x} = 0$ \qquad (8.23\text{-}d)

At $y = 0$, $\tau < 0$ $(0 < x < B/D_h)$: $\dfrac{\partial \theta_s}{\partial y} = 0$ \qquad (8.23\text{-}e)

At $y = 0$, $\tau \geq 0$ $(0 < x < B_u/D_h)$: $\dfrac{\partial \theta_s}{\partial y} = 0$ \qquad (8.23\text{-}f)

At $B_u/D_h < x < B/D_h$: $\dfrac{\partial \theta_s}{\partial y} = -\dfrac{k_f}{k_s}\dfrac{D_h}{B}$ \qquad (8.23\text{-}g)

At $y = H/D_h$: $u = v = w = 0$, $\dfrac{\partial \theta_f}{\partial y} = 0$ (in the fluid region)

$$\frac{\partial \theta_s}{\partial y} = 0 \text{ (in the solid region)} \qquad (8.23\text{-}h)$$

At $\tau < 0$: $\theta_s = \theta_f = 0$, and at the solid-fluid interface (n-normal direction)

$$u = v = w = 0, \quad \theta_f = \theta_s, \quad \frac{\partial \theta_f}{\partial n} = \frac{k_s}{k_f}\frac{\partial \theta_s}{\partial n} \qquad (8.23\text{-}i)$$

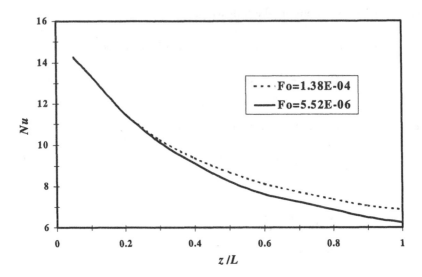

Figure 8.17: Local Nusselt number along the microchannel heat sink at two time intervals [176].

The numerical computations were carried out by solving the equations for the solid and fluid regions simultaneously as a single domain problem with the corresponding boundary conditions. The distribution of the peripheral averaged Nusselt number along the channel is illustrated in Figure 8.17 at two different time intervals. The thermal performance of the microchannel heat sink near the entrance remains approximately constant during the transient, while it is slightly improved with time at locations downstream. The small increase in Nusselt number during the transient is due to diffusion limited heat transfer in the substrate.

The effects of various parameters on the system transient performance have been investigated. Increasing the channel depth/width while keeping a constant flow rate, by reducing the inlet velocity, increases the transient duration. This is due to the higher energy storage capacity of the system, which lengthens its thermal lag. When the Reynolds number is kept constant, the mass flow rate increases with channel width/depth. A larger width/depth of the channel also results in more energy storage capacity of the system and, thus, tends to lengthen the thermal lag in the system. However, a global improvement in the convective heat transfer performance is obtained due to the larger width of the bottom fluid-solid interface. This is the predominant effect that shortens the thermal lag of the system.

Increasing the channel depth while maintaining a constant inlet velocity results in increasing Reynolds number. In this case, the effect of higher energy storage capacity of the system is offset by the effect of the higher

Reynolds number, which reduces the transient period. Consequently, the transient duration remains unaffected. On the other hand, the fluid mean temperature at the exit is significantly reduced because the mass flow rate is higher for a deeper channel with a constant inlet velocity. When decreasing the Reynolds number, by decreasing the fluid velocity for a given channel geometry, the time required to reach steady state increases due to the lower convection heat transfer rate at the solid-fluid interface. The fluid mean temperature at the exit also increases because of the lower mass flow rate.

The transient temperature, for all calculated cases, is found to decrease monotonically for the shutdown of the heat source, similar to the measured response plotted in Figure 8.6-b. However, the temperature increases monotonically also for the start-up of the heat source, and there is no case where an overshoot is observed as evident in Figure 8.6-a. In the calculations, the mixed mean temperature of the working fluid at the channel exit is selected for the analysis for the transient response, while only the surface temperature of the solid device is measured in the experiment. This may or may not be the reason for the difference in the transient response of the system during its heating up process due to a step-power input.

Chapter 9

Micro Heat Pipes

The concept of combining phase-change heat transfer and micro-fabrication techniques to construct micro heat pipes for the dissipation of heat is due to Cotter [34]. Since the introduction of this idea, the proposed applications of micro heat pipes have expanded from the thermal control of localized heat generating devices such as laser diodes and infrared detectors to the removal of heat from the leading edges of stator vanes in turbines or the leading edges of hypersonic aircrafts. While not all the suggested applications have materialized, micro heat pipes have been fabricated, modelled and analysed. The larger of these systems have been implemented in commercially available products such as laptop computers or high-precision equipment. Indeed, most of the research work is still devoted to cooling of semiconductors chips. Micro heat pipes are used to eliminate hot spots, reduce temperature gradients, and improve chip reliability. The high power level currently possible, on the order of 100W/cm^2, together with the heat pipes being self-contained and self-starting make micro heat pipes an ideal heat transfer system. Furthermore, the introduction of newly developed CMOS-compatible techniques to integrate such thermal microsystems with microelectronic devices will surely enhance their range of applications.

Much has been written on this subject, and a number of reviews have previously summarized the early work up to the mid 1990s [164]. However, significant progress has been made over the past few years, particularly in the development of a better understanding of the thin-film behaviour that governs the operation of these micro heat pipes. Peterson has recently summarized in a detailed, up-to-date review the advances made in both individual and arrays of micro heat pipes as well investigations of flat-plate microscale heat spreaders [160].

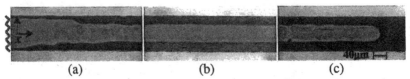

(a)   (b)   (c)

Figure 9.1: The flow pattern at the (a) evaporator, (b) adiabatic, and (c) condenser region during a micro heat pipe operation; the heater is located near the evaporator [109].

9.1   Micro heat pipe operation principle

Micro heat pipes are generally considered to have a hydraulic radius, defined as twice the cross-sectional area divided by the wetted perimeter, comparable to the reciprocal of the mean curvature of the liquid-vapour interface. A dimensionless expression that more clearly defines a micro heat pipe is given by [159]:

$$\frac{r_c}{r_h} \geq 1 \qquad\qquad\qquad (9.1)$$

where r_c and r_h are the capillary and hydraulic radii respectively. Practically, a micro heat pipe consists of a small noncircular channel that utilizes the sharp angled corner regions as liquid arteries.

The fundamental operating principles of micro heat pipes are essentially the same as those of the larger, more conventional heat pipes. Heat applied to one end of the heat pipe vaporizes the liquid in that region and forces it to move to the cooler end where it condenses and gives up the latent heat of vaporization. This vaporization and condensation process causes the liquid-vapour interface in the corner regions to change continually along the pipe, resulting in a capillary pressure difference between the evaporator and condenser regions. This capillary pressure difference promotes the flow of the working fluid from the condenser back to the evaporator through the triangular shaped corner regions. These corner regions in micro heat pipes serve as the liquid arteries, and no additional wicking structure is required. If the ratio between the heat pipe length and its hydraulic diameter is large enough, the working fluid in the pipe undergoes evaporation and condensation with an intermediate adiabatic zone as shown in Figure 9.1.

Significant differences exist between various individual and arrays of micro heat pipes in design parameters and operational conditions. In an attempt to quantify the performance of micro heat pipes, using Fourier's law, an effective thermal conductivity k_{eff} is defined as follows:

$$q = k_{eff} A_{eff} \Delta T / L_{eff} \qquad (9.2)$$

where q is the input power, A_{eff} is the cross-sectional area of the device, ΔT is the difference between the evaporator and condenser average temperatures, and L_{eff} is the effective heat pipe length defined as:

$$L_{eff} = 0.5L_e + L_a + 0.5L_c \qquad (9.3)$$

where L_e, L_a and L_c are the lengths of the evaporator, adiabatic and condenser regions, respectively.

9.2 Maximum heat transfer limitations

The operation and performance of heat pipes are dependent on many factors such as the pipe shape, working fluid and the wick structure. A number of fundamental mechanisms limit the maximum heat transfer rate for heat pipes operating in steady state, which include the capillary limit, viscous limit, sonic limit, entrainment and boiling limits, while the transient operation and start-up dynamics determine the steady state operation.

9.2.1 Capillary limitation

The primary mechanism by which micro heat pipes operate is the result of the difference in the capillary pressure across the liquid-vapour interfaces in the evaporator and condenser. For proper operation, the capillary pressure difference must exceed the sum of all the pressure losses throughout the liquid and vapour flow paths. Body forces, such as gravity, are negligible in microsystems compared to surface forces and, consequently, the hydrostatic pressure drop can be neglected. Therefore, the capillary limit can be expressed as:

$$\Delta P_{mc} \geq \Delta P_f + \Delta P_g \qquad (9.4)$$

where ΔP_{mc} is the net capillary pressure difference, ΔP_f is viscous pressure drop in the liquid phase, and ΔP_g is the viscous pressure drop in the vapour phase. As long as this condition is met, liquid will flow back from the condenser to the evaporator. For cases in micro heat pipes where the summation of the viscous pressure losses is greater than the capillary pressure difference between the evaporator and condenser, the corners are depleted of liquid and dry out. This is the capillary limitation, which depends on the wicking structure, working fluid properties, evaporator heat flux and operating temperature.

The capillary pressure difference at a liquid-vapour interface is given by the LaPlace-Young equation, and for most heat pipe applications it can be reduced to:

$$\Delta P_{mc} = \left(\frac{2\sigma}{r_{ce}} \right) - \left(\frac{2\sigma}{r_{cc}} \right) \qquad (9.5)$$

where r_{ce} and r_{cc} are the radii of curvature in the evaporator and condenser regions, respectively, while σ is the surface tension coefficient. During normal operation of the heat pipe, the vaporization causes the liquid meniscus to recede into the evaporator corners reducing the local capillary radius r_{ce}. In the condenser, on the other hand, condensation increases the local capillary radius r_{cc}. This difference between the two radii of curvature is the driving force for the flow of liquid from the condenser to the evaporator. It is generally assumed that the steady-state capillary radius in the condenser approaches infinity, such that the maximum capillary pressure for a heat pipe can be expressed as a function of only the capillary radius in the evaporator,

$$\Delta P_{mc} \cong \left(\frac{2\sigma}{r_{ce}} \right) \qquad (9.6)$$

Values for effective capillary radius can be estimated theoretically for simple geometries or experimentally for more complicated structures [24].

The liquid flows back from the condenser to the evaporator against viscous forces, which can be written in terms of frictional drag:

$$\frac{dP_f}{dx} = -\frac{2\tau_f}{r_{hf}} \qquad (9.7)$$

where τ_f is the frictional shear stress, primarily at the liquid-solid interface, and r_{hf} is the hydraulic radius. For steady-state operation with constant heat addition and removal, the equation can be integrated over the length of the micro heat pipe to yield the liquid pressure drop:

$$\Delta P_f = \left(\frac{\mu_f}{KA_f h_{fg} \rho_f} \right) L_{eff} q \qquad (9.8)$$

μ_f, ρ_f and h_{fg} are the liquid viscosity, density and latent heat of vaporization, respectively, while K and A_f are the wick permeability and the wick cross-sectional area.

The vapour pressure drop can be estimated similarly to the liquid pressure drop, but it is more complicated due to the mass addition and removal in the evaporator and condenser, respectively, as well as the compressibility of the vapour flow. Consequently, a more accurate estimate has to include the dynamic pressure. In-depth analysis for determining the overall vapour pressure drop has been presented by several authors [159]. The resulting expression, for practical values of Reynolds and Mach numbers, is similar to the liquid pressure drop:

$$\Delta P_g = \left(\frac{16\mu_g}{2r_{hg}^2 A_g h_{fg} \rho_g} \right) L_{eff} q \tag{9.9}$$

μ_g and ρ_g are the vapour viscosity and density, respectively, while r_{hg} and A_g are the hydraulic radius and cross-sectional area of the vapour space.

9.2.2 Viscous limitation

At a very low heat pipe operating temperature, the vapour pressure difference between the evaporator, the high-pressure region, and the condenser, the low-pressure region, can be extremely small. The viscous forces within the vapour region may then prove to be dominant, if the pressure difference is too small, and hence limit the heat pipe operation. The following criterion [39],

$$\frac{\Delta P_g}{P_g} < 0.1 \tag{9.10}$$

has been suggested for determining when this limit might be of a concern. For either steady-state operation or applications with moderate operating temperature range, the viscous limitation will not be important.

9.2.3 Sonic limitation

The sonic limitation in heat pipes is the result of vapour velocity variations along the heat pipe due to the axial variations in the vaporization and condensation processes. Similar to the effect of decreased outlet pressure in a converging–diverging nozzle, decreased condenser temperature leads to a decrease in the evaporator temperature, up to but not beyond that point where choked flow occurs in the evaporator; hence, reaching the sonic limit. Any further decrease in the condenser temperature does not reduce the evaporator temperature or the maximum heat transfer capability, due to the development of choked flow. The sonic limitation of maximum heat flux in heat pipes, q_{ms}, is given by [24]:

$$q_{ms} = A_g \rho_g h_{fg} \left(\frac{\gamma_g R_g T_g}{2\gamma_g + 2} \right)^{1/2} \tag{9.11}$$

where R_g, γ_g and T_g are the vapour gas constant, ratio of specific heats, and mean temperature within the heat pipe, respectively.

9.2.4   Entrainment limitation

During heat pipe operation, the liquid and the vapour flow in opposite directions, resulting in a shear stress at the interface. At very high heat fluxes, liquid droplets could be picked up or entrained into the vapour flow. This entrainment may result in dryout of the evaporator wick due to excess liquid accumulated in the condenser. The Weber number We, representing the ratio of the viscous shear force to the liquid surface tension force, defined as:

$$We = \frac{2r_{hf} \rho_g V_g^2}{\sigma} \tag{9.12}$$

can be used to determine at which condition this entrainment is likely to occur. To prevent the entrainment of liquid droplets into the vapour flow, the Weber number must therefore be less than one. This implies that the maximum transport capacity based on the entrainment limitation, q_{me}, can be expressed as follows [39]:

$$q_{me} = A_g h_{fg} \left(\frac{\sigma \rho_g}{2r_{hf}} \right)^{1/2} \tag{9.13}$$

where r_{hf} is the hydraulic radius of the wick structure.

9.2.5   Boiling limitation

All the limits discussed previously depend upon the axial heat transfer. In contrast, the boiling limit depends upon the evaporator heat flux. It occurs when nucleate boiling in the evaporator wick generates vapour bubbles that partially block the return of fluid. The presence of vapour bubbles in the wick requires both the formation of bubbles and also the subsequent growth of these bubbles. The maximum heat flux due to the boiling limit, q_{mb}, can be written as [24]:

$$q_{mb} = \left(\frac{2\pi L_{eff} k_{eff} T_g}{h_{fg} \rho_g \ln(r_i / r_v)} \right) \left(\frac{2\sigma}{r_n} - \Delta P_{mc} \right)$$ (9.14)

where r_i is the inner radius of the heat pipe wall, and r_n is the nucleation site radius [39].

9.3   Individual micro heat pipes

The early micro heat pipes typically consisted of a long thin tube with one or more small noncircular channels that utilized the sharp-angled corner regions as the wicking structure. Analytical and experimental investigations have been conducted to explore both the steady state and the transient performance of micro heat pipes.

9.3.1   Steady state operation

The first steady-state analytical models of individual micro heat pipes utilized the traditional pressure balance approach developed for use in more conventional heat pipes. These models provided a mechanism by which the steady-state and transient performance characteristics of micro heat pipes could be determined and indicated that, while the operation was similar to that observed in larger more conventional heat pipes, the relative importance of many of the parameters is quite different. Perhaps the most significant difference was the relative sensitivity of the micro heat pipes to the amount of working fluid present.

The capillary limit has been shown to be the limiting factor in almost all types of micro heat pipes. The vapour flow can have a significant effect on the liquid flow in the micro heat pipe, although the vapour pressure drop is very small compared with the liquid pressure drop. Under uniform heat flux at the evaporator and condenser sections, the capillary flow in the triangular capillary grooves is governed by the capillary force; hence, neglecting gravitational effects, the pressure drop produced by the shear forces at the solid-liquid and liquid-vapour interfaces must be overcome by the capillary pumping force. The net capillary pressure difference can be estimated from Equation 9.5.

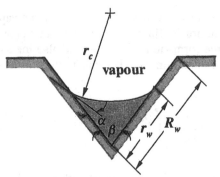

Figure 9.2: The flow structure in the triangular groove showing the interface between the liquid and vapour zones.

For capillary flows at very low Bond numbers, the surface of the liquid will have a nearly constant radius of curvature, which depends on the geometric shape of the liquid flow passage and the contact angle. Based on this assumption, the length of the wetted wall r_w, illustrated in Figure 9.2, can be expressed in terms of the radius of curvature r_c as follows:

$$r_w = r_c \frac{\cos(\alpha + \beta/2)}{\sin(\beta/2)} \tag{9.15}$$

where β is the channel angle and α is the contact angle, which depends on the working-liquid and solid-surface properties as well as the applied heat flux. Decreasing the capillary radius results in not only increased capillary pressure but also increased frictional pressure drop. However, the rate at which the frictional force increases is larger than the increasing rate of the capillary force. Consequently, a lower bound on the capillary radius exists, namely a minimum capillary radius.

In the triangular groove, the capillary flow of an incompressible Newtonian fluid is considered with fully developed laminar vapour flow over the liquid-vapour interface as illustrated in Figure 9.2. It is clear that the flow velocity and behaviour of the liquid surface will be strongly influenced by the vapour flow direction and velocity. Utilizing a cylindrical coordinate system (r, φ) with its origin coincident with the triangular apex and assuming two-dimensional fully developed liquid flow, the pressure drop in the triangular groove can be expressed as:

$$\frac{1}{r} \frac{\partial}{\partial r} \left(r \frac{\partial u_f}{\partial r} \right) + \frac{1}{r^2} \frac{\partial^2 u_f}{\partial \varphi^2} = \frac{1}{\mu_f} \frac{dp_f}{dx} \tag{9.16}$$

The boundary conditions are given by:

$$u_f = 0 \quad @ \quad \varphi = 0 \text{ and } \varphi = \beta \qquad (9.17\text{-a})$$

$$u_f = u_{ft} \quad @ \quad r = r_t \qquad (9.17\text{-b})$$

where r_t is the location of the liquid-vapour interface with respect to the apex. The heat transfer through the liquid flow in the triangular groove can be approximated by the two-dimensional heat conduction equation:

$$\frac{\partial^2 T}{\partial r^2} + \frac{1}{r}\frac{\partial T}{\partial r} + \frac{1}{r^2}\frac{\partial^2 T}{\partial \varphi^2} = 0 \qquad (9.18)$$

Since the heat conductivity of the wall is much larger than the liquid conductivity, the temperature of the wall at each cross-section can be assumed to be uniform. Hence, the boundary conditions are given by:

$$T = T_w \quad @ \quad \varphi = 0 \text{ and } \varphi = \beta \qquad (9.19\text{-a})$$

$$T = T_{fg} \quad @ \quad r = r_t \qquad (9.19\text{-b})$$

For the vapour flow, it is very difficult to obtain the boundary conditions since the vapour space shape is not regular and evaporation occurs near the interline region. Therefore, a simplified model has been suggested where the pressure drop is determined at each axial location by the following two-dimensional model:

$$\frac{\partial^2 u_g}{\partial y^2} + \frac{\partial^2 u_g}{\partial z^2} = \frac{1}{\mu_g}\frac{dp_g}{dx} \qquad (9.20)$$

After obtaining the friction factor based on the cross-sectional shape of the flow path, the one-dimensional momentum equation for the vapour axial flow is expressed as

$$\frac{dp_g}{dx} + \rho_g u_{ag}\frac{du_g}{dx} = -f_g \frac{2\rho_g u_{ag}^2}{D_{hg}} \qquad (9.21)$$

where u_{ag} is the vapour average velocity, and D_{hg} is the hydraulic diameter of the vapour space.

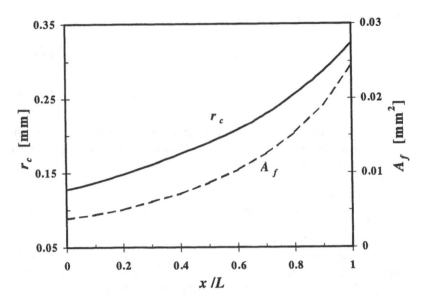

Figure 9.3: Variations of the meniscus radius and the cross-sectional area of the liquid flow as a function of the distance from the evaporator [162].

The corresponding temperature variation can be estimated based on the Clapeyron equation as follows:

$$T_g = \frac{T_{g0}}{1 - \left(RT_{g0} / h_{fg}\right)\ln\left(p_g / p_{g0}\right)}$$
(9.22)

T_{g0} and p_{g0} are the reference temperature and pressure, respectively.

Peterson and Ma [162] computed the variations of the meniscus radius and cross-sectional area of the liquid flow in the triangular groove with apex angle of 60° as shown in Figure 9.3. Both the cross-sectional area of the liquid flow A_f as well as the meniscus radius r_c increase gradually along the heat pipe from the evaporator to the condenser.

The most basic parameters affecting the performance of a micro heat pipe are its size and shape. In the case of micro heat pipes, the solid substrate has a large thermal mass, relative to the size of the heat pipe, and the system cannot be considered isothermal. Therefore, a simple one-dimensional conjugate model of a micro heat pipe during steady state operation has been developed to assess the size effect accounting for the heat conduction [114]. This model uses the predicted maximum heat, at capillary limitation, and determines the total required heat input to overcome the heat conduction in the solid substrate.

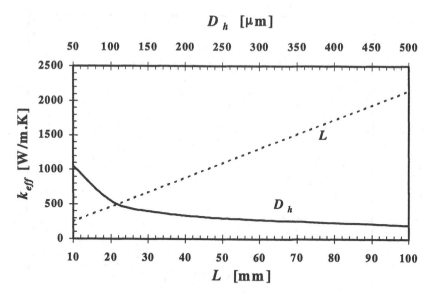

Figure 9.4: The dependence of the effective thermal conductivity of a micro heat pipe on its length and its hydraulic diameter [114].

The model, which is basically a thermal resistance model, predicts the effective thermal conductivity k_{eff} of the substrate/working fluid micro heat pipe as illustrated in Figure 9.4. The effective thermal conductivity is directly proportional to the pipe length, L, and inversely proportional to the width or hydraulic diameter, D_h, of the pipe.

The micro heat pipe operation is sensitive to the fill quantity of the working fluid. For a conventional heat pipe, the fill quantity is usually less than 10% of the total volume. Hence one would expect the same or even less fill quantity for a micro heat pipe, as the size of a heat pipe is orders of magnitude smaller than a conventional pipe. Duncan and Peterson [38] theoretically investigated the charge optimisation for a triangular-shaped micro heat pipe. Using the information provided by the capillary limit analysis, the radius of curvature in the evaporator region of the micro heat pipe was calculated for all input power values less than the capillary limit. With the known radius of curvature in the evaporator, adiabatic and condenser regions, the optimal liquid charge was calculated as a function of input power, and the results are plotted in Figure 9.4. The resulting optimal liquid charge, χ_{op}, decreases from about 25% of the total micro heat pipe volume at input power values approaching zero to approximately 16% at the capillary limit of the micro heat pipe. Hence, the optimal charge value does not exceed 25% even for input levels near zero.

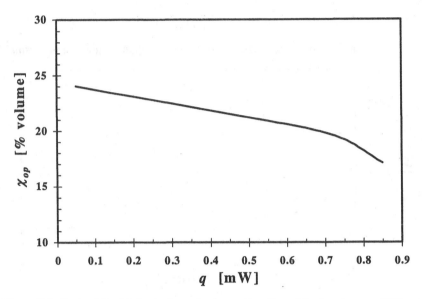

Figure 9.5: Optimal liquid charge by volume as a function of the input power [38].

As fabrication capabilities have developed, experimental studies of individual micro heat pipes have been conducted on progressively smaller devices, beginning with early investigations on mini heat pipes, about 3mm in diameter, and progressing to micro heat pipes in the 30-μm-diameter range. These investigations have included both steady state and transient investigations. In the earliest experimental tests [5], several micro heat pipes approximately 1 mm in diameter were evaluated. The results of this work represented the first successful operation of a mini/micro heat pipe that utilized the principles outlined in the original concept of Cotter [34], and as such paved the way for numerous other investigations and applications.

The steady-state performance limitations and operational characteristics of a trapezoidal heat pipe were predicted reasonably well in comparison with experimental results for operating temperatures between 40 and 60°C [5]. The maximum heat transport capacity of the micro heat pipe is shown in Figure 9.6 as a function of the operating temperature. The steady-state model over-predicts the measured heat transport capacity at operating temperatures below 40°C, and under-predicts it at operating temperatures above 60°C. The experimentally measured steady-state temperature distribution is compared in Figure 9.7 with the distribution predicted by the numerical model for a power level of 0.12W [230]. Throughout the entire pipe length, the predicted temperature distribution is very close to the measured distribution.

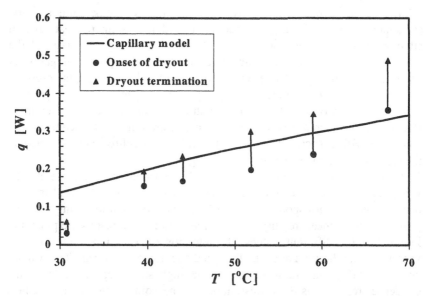

Figure 9.6: A comparison between measured and calculated maximum heat transport capacity of a trapezoidal micro heat pipe as a function of the operating temperature [5].

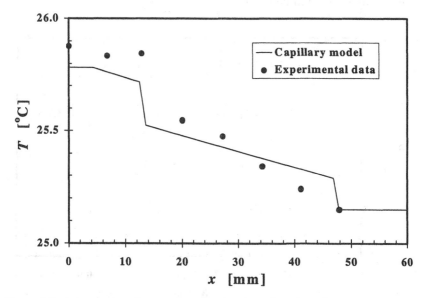

Figure 9.7: A comparison between measured and predicted steady-state temperature distribution for an input power of 0.12W [230].

In a recent experimental work [109], alongside temperature sensors, capacitance sensors have also been integrated into a micro heat pipe in attempt to measure the void fraction distribution taking advantage of the large difference between the dielectric constants of liquids and gases. A possible calibration procedure, which turned out to be difficult, is described in detail in Chapter 5. Furthermore, the sensor capacitance is reported to increase with working liquid temperature, decrease with measurement frequency, and increase with liquid ion concentration. These trends are not consistent with expectations, and could be due to the direct contact between the capacitor electrode and the working liquid.

At any rate, with a reliable calibration curve accounting for these effects, void fraction measurements are still possible. Indeed, void-fraction distributions v, corresponding to the liquid-vapour mixture along the heat pipe, have been measured using capacitive sensors. The results are plotted in Figure 9.8 for different values of input power. The transition from the vapour to the liquid zone via the adiabatic region is clear. The vapour content in the evaporator region is the highest, about 1, while in the condenser region it is the lowest, about 0. The void fraction along the heat pipe increases with input power. Namely, the extent of the condenser region decreases while the extent of the adiabatic region increases with increasing input power. These expected results are consistent with the recorded flow visualizations shown in Figure 9.1.

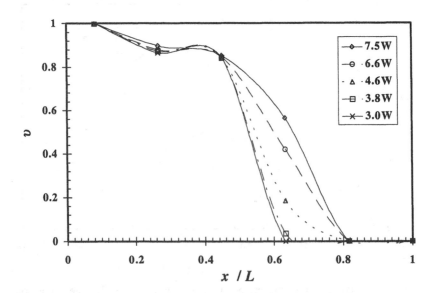

Figure 9.8: Void-fraction distribution measurements along the micro heat pipe, from evaporator $x/L=0$ to condenser $x/L=1$, for different input-power level [109].

9.3.2 Transient performance

As heat pipes diminish in size, the transient nature becomes of increasing interest. The ability to respond to rapid changes in heat flux coupled with the need to maintain constant evaporator temperature in modern high-powered electronics necessitates a complete understanding of the temporal behavior of these devices. The first reported transient investigation of micro heat pipes was conducted by Wu and Peterson [229]. The most interesting result from this model was the observation that reverse liquid flow occurred during the startup of micro heat pipes. This reverse liquid flow is the result of an imbalance in the total pressure drop and occurs because the evaporation rate does not provide an adequate change in the liquid–vapor interfacial curvature to compensate for the pressure drop. As a result, the increased pressure in the evaporator causes the meniscus to recede into the corner regions, forcing liquid out of the evaporator and into the condenser. During startup, the pressure of both the liquid and vapor is higher in the evaporator and gradually decreases with position, promoting flow away from the evaporator. Once the heat input reaches full load, the reverse liquid flow disappears and the liquid mass flow rate into the evaporator gradually increases until a steady-state condition is reached. At this time, the change in the liquid mass flow rate is equal to the change in the vapor mass flow rate for any given section.

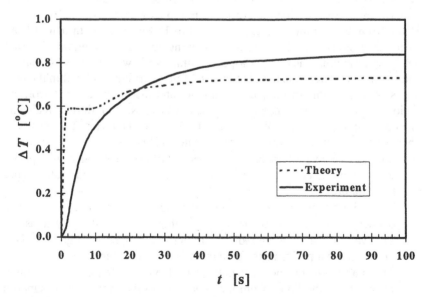

Figure 9.9: A comparison between measured and predicted transient response for a tapered micro heat pipe [159].

A comparison of the predicted results with the experimental data indicated the importance of the liquid charge, the contact angle and the shear stresses at the liquid–vapor interface in predicting the maximum heat-transfer capacity and thermal resistance of these devices. The time dependent temperature distributions were measured to determine the transient response characteristics of the heat pipe as a function of incremental power increase and mean operating temperature. The measured and calculated temperature difference between a micro heat pipe evaporator and condenser, ΔT, are compared in Figure 9.9. The transient model predicts a much faster response than the measured response. However, with time, the measured temperature difference approaches and exceeds the predicted value by approximately 15%. The initial deviation of the actual response from that predicted could in part be due the time constant associated with the heater. The effect of the reversed liquid flow observed in the calculations, the slight inflection at $t=12$s, is not apparent in the experimental results.

Several other transient analyses have been proposed [160]. Khrustalev and Faghri [94] presented a detailed mathematical model of the heat and mass transfer processes in micro heat pipes describing the distribution of the liquid and the thermal characteristics as a function of the liquid charge. The liquid flow in the triangular-shaped corners of a micro heat pipe with a polygonal cross section was considered by accounting for the variation of the curvature of the free liquid surface and the interfacial shear stresses due to the liquid–vapor interaction. Itoh and Polásek [73] presented the results of an extensive experimental investigation on a series of micro heat pipes ranging in size and shape from 1 to 3 mm in diameter and 30 to 150 mm in length that utilized different cross-sectional configurations along with a conventional internal wicking structure. The unique aspect of this work was the use of neutron radiography to determine the distribution of the working fluid within the heat pipes. Applying this technique, the amount and distribution of the working fluid as well as non-condensable gases were observed during real-time operation along with the boiling and/or re-flux flow behavior. The results of these tests indicated several important results [158]:

- Similar to conventional heat pipes, the maximum heat-transport capacity of a micro heat pipe mainly depends on the mean adiabatic vapor temperature.
- Micro heat pipes having smooth inner surfaces are more sensitive to overheating compared to micro pipes with grooved capillary systems.
- The wall thickness of the individual micro heat pipes has a greater effect on the thermal performance than the pipe casing material.
- The maximum transport capacity of heat pipes utilizing axial channels for return of the liquid from the condenser to the evaporator is superior to that of heat pipes with a formal wicking structure.

9.4   Arrays of micro heat pipes

The initial concept of micro heat pipes by Cotter (1984) envisioned fabricating micro heat pipes directly into semiconductor devices. In microsystem technology, this is a natural consequence of batch fabrication since the construction of an array of micro heat pipes requires the same effort (cost) as the construction a single heat pipe. While many of the analytical models developed can be used to predict the performance limitations and operational characteristics of individual micro heat pipes, it is not clear how the incorporation of an array of these devices might affect the temperature distribution and the resulting thermal performance.

9.4.1   Steady state modelling and testing

Mallik *et al.* [125] developed a three-dimensional numerical model capable of predicting the thermal performance of an array of parallel micro heat pipes constructed as an integral part of semiconductor chips. In order to determine the potential advantages of this concept, several different thermal loading configurations were analyzed. The reduction in maximum surface temperature, the mean chip temperature and the maximum temperature gradient across the chip as a function of the number of micro heat pipes in the array was determined.

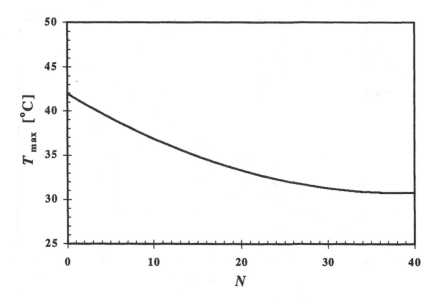

Figure 9.10: The effect of the number of micro heat pipes in an array on the maximum chip surface temperature [125].

The effect of variations in the number of heat pipes in an array, N, on the thermal performance is clearly demonstrated in Figure 9.10. Increasing the number of heat pipes can significantly decrease the maximum chip surface temperature, T_{max}. Although, as expected, the maximum chip temperature decreases monotonically (but not linearly) with an increasing number of pipes, the relative decrease drops off significantly around 19 heat pipes.

Steady-state thermal behavior of arrays of micro heat pipe was investigated experimentally [123]. The test devices included silicon dies, each $8 \times 20 mm^2$ in area and 0.378mm in thickness, with and without triangular micro heat pipes, 25μm wide and 55μm deep. The variations in the maximum surface temperature difference ΔT_s as a function of the input power are shown in Figure 9.11. Clearly, the surface temperature gradient is highest for the die with no heat pipes. The temperature gradient decreases significantly for a die with an array of 34 micro heat pipes, occupying 0.7% of the die cross-sectional area, and a bit less for the array of 66 micro heat pipes, occupying 1.4% of the cross-sectional area. The effective thermal conductivity of the three dies is compared in Figure 9.12 as a function of the input power. No measurable difference could be detected between the two array densities. However, a significant difference in both the magnitude and slope of the conductivity of the die with no heat pipes is evident. For a die with no heat pipes, the low thermal conductivity decreases with input power; while with the array, the high conductivity increases with input power.

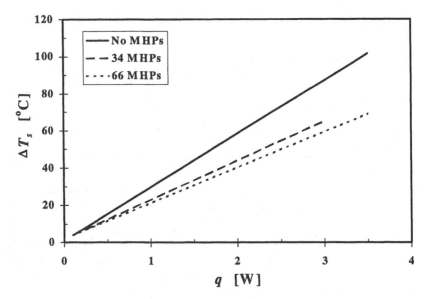

Figure 9.11: The maximum surface temperature difference as a function of the input power for chips with and without a micro heat pipe array [123].

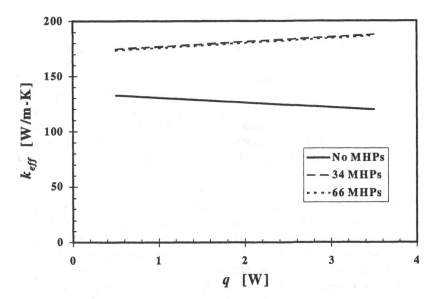

Figure 9.12: The effective thermal conductivity as a function of the input power for chips with and without a micro heat pipe array [123].

The effect of the micro heat pipe shape on the array thermal performance was also studied experimentally [161], and the results are summarized in Figure 9.13. As has been demonstrated, the maximum surface temperature of the die with the no micro heat pipes array is significantly higher. Moreover, the die with the triangular heat pipes has the lowest maximum temperature over the entire tested range of input power (Figure 9.13-a). As the surface temperature decreases the thermal conductivity increases (Figure 9.13-b); thus, the thermal conductivity of triangular micro heat pipes is the highest.

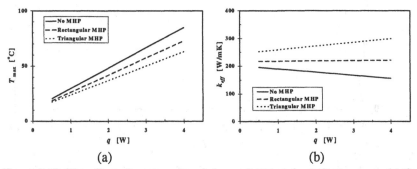

(a)   (b)

Figure 9.13: The effect of cross-sectional shape of a micro heat pipe array on: (a) the maximum die temperature, and (b) the effective thermal conductivity as a function of the input power [161].

9.4.2   Transient modelling and testing

The three-dimensional numerical model of Mallik *et al.* [125] was further extended to determine transient response characteristics of an array of micro heat pipes fabricated into silicon wafers. The numerical simulations were used to predict the time-dependent temperature distribution occurring within the wafer, given the physical parameters of the wafer and the locations of the heat sources and sinks. The results indicated that significant reductions in the maximum localized wafer temperatures and thermal gradients across the wafer could be obtained through the incorporation of an array of micro heat pipe.

The transient thermal response was measured and compared with the calculations based on the numerical model [163]. Dies with arrays of either 34 or 66 micro heat pipes were evaluated using an infrared thermal imaging system in conjunction with a VHS video recorder. These arrays occupied 0.75% and 1.45% of the wafer cross-sectional area, respectively. To better represent the transient thermal response of the different devices, the non-dimensional temperature, $\theta=(T-T_i)/(T_{ss}-T_i)$, and time, t/τ, are used. T_i is the initial wafer temperature, and τ is the time constant. T_{ss} is the steady-state temperature, which depends not only on the input power but also on the location of the point.

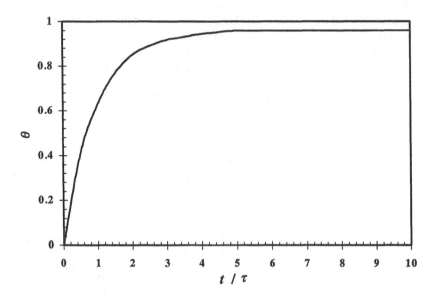

Figure 9.14: Dimensionless temperature as a function of time for dies with and without micro heat pipe array [163].

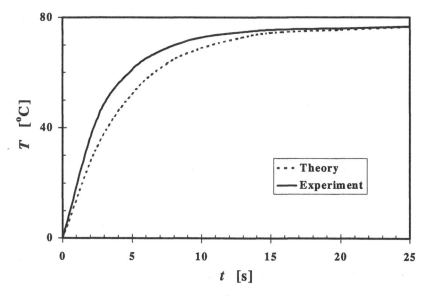

Figure 9.15: A comparison between predicted and measured transient temperature response for a die with an array of 66 micro heat pipes [163].

Using the nondimensional variables, the measured time response for the dies with and without micro heat pipes collapse together into a single curve shown in Figure 9.14. The resulting response curve can be expressed as:

$$\theta = 1 - \exp(-t/\tau) \qquad\qquad (9.23)$$

Similar to microchannel heat sinks (dry device), this is a typical response of a first order system. A comparison between measured and calculated transient response for a die with an array of 66 micro heat pipes is shown in Figure 9.15. The agreement between the two curves is quite close over the entire time domain with the largest variation occurring in the intermediate range, i.e. the range where the time constant is calculated. The steady-state values are about the same with no noticeable difference demonstrating the accuracy of the numerical model. The results showed a slight increase in the calculated time constant with increasing input power for the dies with and large increase for the die without micro heat pipes. The reduction in the time constant for dies with micro heat pipe arrays was 30-45% over the plain die. This is a direct result of the increase in the effective thermal conductivity, caused by the vaporization and condensation with the heat pipes. Finally, using micro heat pipes leads to a significant decrease in the maximum die temperature for a given power input, as well as a significant reduction in the transient thermal gradients due to the improved response time.

References

1. Adams, T.M., Abdel-Khalik, S.I., Jeter, S.M. and Qureshi, Z.H., An experimental investigation of single-phase forced convection in microchannels, *Int. J Heat Mass Transfer*, vol.41, pp.851–857, 1998.

2. Ajay, G., Richard, C.J., Sushil H.B., Charles, D.E., Narendra, K.P., Mehdi, A. and John, S.G., Formation of silicon reentrant cavity heat sinks using anisotropic etching and direct wafer bonding, *IEEE Elctron Device Letters*, vol.14, pp.29-32, 1993.

3. Arkilic, E.B., Breuer, K.S. and Schmidt, M.A., Gaseous flow in microchannels, in *Application of Microfabrication to Fluid Mechanics*, ASME FED-vol.197, pp.57-66, 1994.

4. Arkilic, E.B., Schmidt, M.A. and Breuer, K.S., Gaseous slip flow in long microchannels, *J. Microelectromech. Syst.*, vol.6, pp.167-178, 1997.

5. Babin, B.R., Peterson, G.P. and Wu, D., Steady-state modeling and testing of a micro heat pipe, *J. Heat Transfer*, vol.112, pp.595–601, 1990.

6. Bang, D.S., Cao, M., Wang, A., Saraswat, K.C. and King, T.J., Resistivity of boron and phosphorus doped polycrystalline Si/sub 1-x/Ge/sub x/ films, *Applied Physics Letters*, vol.66, pp.195-197, 1995.

7. Bankoff, S.G., Entrapment of gas in the spreading of liquid over a rough surface, *AIChE J.*, vol.4, pp.24-26, 1958.

8. Bau, H. H., Optimization of conduits' shape in microscale heat exchangers, *Int. J. Heat Mass Transfer*, vol.41, pp. 717–2723, 1998.

9. Baxter, L.K., *Capacitive Sensors: design and applications*, IEEE Press, New York, 1997.

10. Belgrader, P., Benett, W., Hadley, D., Long, G., Mariella, R., Jr., Milanovich, F., Nasarabadi, S., Nelson, W., Richards, J. and Stratton, P., Rapid pathogen detection using a microchip PCR array instrument, *Clin. Chem.*, vol.44, pp.2191-2194, 1998.

11. Berg, H.R. van den, Seldam, C.A. ten and Gulik, P.S. van der, Compressible laminar flow in a capillary, *J. Fluid Mech.*, vol.246, pp.1-20, 1993.

180

12. Berlicki, T.M., Murawski, E., Muszynski, M., Osadnik S.J. and Prociow, E.L., Thin-film thermocouples of Ge doped with Au and B, *Sensors Actuators A*, vol.50, pp.183-186, 1995.

13. Berlicki, T., Osadnik, S. and Prociow, E., Thermal thin-film sensors for RMS value measurements, *Sensors Actuators A*, vol.27, pp.629-632, 1991.

14. Beskok, A. and Karniadakis, G.E., Simulation of heat and momentum transfer in complex microgeometries, *J. Thermophys. Heat Transfer*, vol.8, pp.647-655, 1994.

15. Beskok A, Karniadakis G.E. and Trimmer W., Rarefaction and compressibility effects in gas microflows, *J. Fluids Eng.*, vol.118, pp.448-456, 1996.

16. Bhatti, M.S. and Shah, R.K., Turbulent and transition flow convective heat transfer in ducts, in *Handbook of Single Phase Convective Heat Transfer*, S. Kakac, R.K. Shah and W. Aung, Eds., Wiley, New York, 1987.

17. Bier, W., Keller, W., Linder, G., Seidel, D., Schubert, K. and Martin H., Gas to gas heat transfer in micro heat exchangers, *Chemical Engineering and Processing*, vol.32, pp.33-43, 1993.

18. Blackwelder, R.F., Hot-wire and hot-film anemometers, *Methods of Experimental physics: Fluid Dynamics*, Emrich R.J. vol.18, Part A, Academic Press, New York-London, 1991.

19. Bordi, F., Cametti, C. and Paradossi, G., Dielectric behavior of polyelectrolyte solutions: the role of proton fluctuation, *J. Phys. Chem.*, vol.95, pp.4883-4889, 1991.

20. Bowers, M.B. and Mudawar, I., High flux boiling in low flow rate, low pressure drop mini-channel and micro-channel heat sinks, *Int. J. Heat Mass Transfer,* vol.37, pp.321-332, 1994.

21. Campbell, D. S. and Hayes, J. A., *Capacitive and Resistive Electronic Components*, Gordon and Breach, New York, 1994.

22. Chaudhari, A.M., Woudenberg, T.M., Albin, M. and Goodson K.E., Transient liquid crystal thermometry of microfabricated PCR vessel arrays, *J. Microelectromech. Syst.*, vol.7, pp.345-355, 1998.

23. Cheng, J., Shoffner, M.A., Hvichia, G.E., Kricha, L.J. and Wilding, P., Chip PCR. II. Investigation of different PCR amplification systems in microfabricated silicon-glass chips, *Nucleic Acids Research*, vol.24, pp.380-385, 1996.

24. Chi, S.W., *Heat Pipe Theory and Practice*, McGraw-Hill, New York, 1976.

25. Choi, S.B., Barron, R.F. and Warrington, R.O., Fluid flow and heat transfer in microtubes, in *Micromech. Sensors, Actuators Syst.*, ASME DSC-vol.32, pp. 123–134, 1991.

26.   Chu, R.K.-H. and Zohar, Y., A class of discrete kinetic solutions for non-boundary-driven gas flow, *J. Non-Equilib. Thermodyn.,* vol.25, pp.49-62, 2000.

27.   Chu, R.K.-H. and Zohar, Y., Non-equilibrium temperature and velocity fields in a microchannel flow using discrete kinetic approach, *J. Non-Equilib. Thermodyn.,* vol.26, pp.15-29, 2001.

28.   Churchill, S.W. and Bernstein, M., A correlating equation for forced convection from gases and liquids to a circular cylinder in crossflow, *J. Heat Transfer,* vol.94, pp.300-306, 1977.

29.   Collier, J.G. and Thome, J.R., *Convective Boiling and Condensation,* Clarendon Press, Oxford, 1994.

30.   Cooper, M.G. and Lloyd, A.J.P., The microlayer in nucleate pool boiling, *Int. J. Heat Mass Transfer,* vol.12, pp.895-913, 1969.

31.   Copeland, D., Behnia, M. and Nakayama, W., Manifold microchannel heat sinks: isothermal analysis, *IEEE Trans. Components, Packaging, Manuf. Technol. A,* vol.20, pp.96–102, 1997.

32.   Corman, T., Enoksson, P. and Stemme, G., Low-pressure-encapsulated resonant structures with integrated electrodes for electrostatic excitation and capacitive detection, *Sensors Actuators A,* vol.66, pp.160-166, 1998.

33.   Cornwell, K. and Brown, R.D., Boiling surface topography, *Proc. 6th Int. Heat Transfer Conf.,* vol.1, pp.157-161, 1978.

34.   Cotter, T.P., Principles and prospects of micro heat pipes, *Proc. 5th Int. Heat Pipe Conf.,* pp.328–335, 1984.

35.   DARPA, Heat Removal by Thermo-Integrated Circuits (HERETIC) /2001, [Online]. Available: http://www.darpa.mil/MTO/heretic/presentations/pres_2001_A/

36.   DeVoe, D.L., Thermal issues in MEMS and microscale systems, in *Thermal Challenges in Next Generation Electronic Systems,* Y.K. Joshi and S.V. Garimella, Eds., Millpress, Rotterdam, pp.25-34, 2002.

37.   Dhir, V.K., Boiling heat transfer, *Annu. Rev. Fluid Mech.,* vol.30, pp.365-401, 1998.

38.   Duncan, A.B. and Peterson, G.P., Charge optimization of triangular shaped micro heat pipes, *J. Thermophys. Heat Transfer,* vol.9, pp.365–368, 1995.

39.   Dunn, P.D. and Reay, D.A., *Heat Pipes, 3rd ed.,* Pergamon Press, New York, 1982.

40.   Eringen, A.C., Theory of micropolar fluids, *J. Math. Mech.,* vol.16, pp.1-18, 1966.

41.   Evans, J., Liepmann, D. and Pisano, A.P., Planar laminar mixer - MEMS fluid processing system, *Proc. 10th Int. Workshop Micro Electro Mechanical Systems, MEMS'97,* pp.96-101, 1997.

182

42. Farrens, S.N., Hunt, C.E., Roberds, B.E. and smith, J.K., A kinetic study of the bonding strength of direct bonded wafers, *J Electrochem. Soc.*, vol.141, pp.3225-3230, 1994.

43. Fleischer, A.S., McAssey, E.V., Jr. and Jones, G.F., Influence of bubble formation on pressure drop in a uniformly heated vertical channel, ASME HTD-vol.342, *National Heat Transfer Conference*, vol.4, pp.51-56, 1997.

44. Friedrich, R. and Patel, C., Towards planetary scale computing - Technical challenges for next generation Internet computing, in *Thermal Challenges in Next Generation Electronic Systems*, Y.K. Joshi and S.V. Garimella, Eds., Millpress, Rotterdam, pp.3-4, 2002.

45. Frohlich, H., *Theory of Dielectric: dielectric constant and dielectric loss, 2nd ed.*, Clarendon Press, Oxford, 1986.

46. Gambill, W.R. and Greene, N.D., Boiling burnout with water in vortex flow, *Chem. Engng. Prog.*, vol.54, pp.68-76, 1958.

47. Gardner J.W., *Microsensors: Principles and Applications*, Wiley, Chichester, 1994.

48. Geraets, J.J.M. and Borst, J.C., A capacitance sensor for two-phase void fraction measurement and flow pattern identification, *Int. J. Multiphase Flow*, vol.14, pp.305-320, 1988.

49. Gillot, C., Meysenc, L. and Schaeffer, C., Integrated single and two phase micro heat sinks under IGBT chips, *IEEE Trans. Components, Packaging, Manuf. Technol. A*, vol.22, pp.384–389, 1999.

50. Goodson, K.E., Kurabaysshi, K. and Pease, R.F.W., Improved heat sinking for laser-diode arrays using microchannels in CVD diamond, *IEEE Trans. Components, Packaging, Manuf. Technol. B*, vol.20, pp.104–109, 1997.

51. Guo, Z.Y. and Wu, X.B., Compressibility effect on the gas flow and heat transfer in a micro tube, *Int. J. Heat Mass Transfer*, vol.40, pp.3251–3254, 1997.

52. Gupta, B., Goodman, R., Jiang, F., Tai, Y.C., Tung, S., Ho, C.M., Analog VLSI system for active drag reduction, *IEEE Micro*, vol.16, pp.53-59, 1996.

53. Hall, D.D. and Mudawar, I., Ultra-high critical heat flux (CHF) for subcooled water flow boiling. II. High-CHF database and design equations, *Int. J. Heat Mass Transfer*, vol.42, pp.1429-1456, 1999.

54. Harendt, C., Graf, H.G., Penteker, E. and Hofflinger, B., Wafer bonding: investigation and in situ observation of the bond process, *Sensors Actuators A*, vol.23, pp.927-930, 1990.

55. Harley, J.C., Huang, Y., Bau, H.H. and Zemel, J.N., Gas flow in micro-channels, *J. Fluid Mech.*, vol.284, pp.257-274, 1995.

56.   Harms, T.M., Kazmierczak, M.J. and Gerner, F.M., Developing convective heat transfer in deep rectangular microchannels, *Int. J. Heat Fluid Flow*, vol.20, pp.149–157, 1999.

57.   Harris, C., Despa, M., and Kelly, K., Design and fabrication of a cross flow micro heat exchanger, *J. Microelectromech. Syst.*, vol.9, pp.502-508, 2000.

58.   Hetsroni, G., private communication, 2002.

59.   Hewitt, G.F., Pheomenological issues in forced convective boiling, in *Convective flow boiling*, J.C. Chen, Ed., Taylor & Francis, Washington, DC, pp.3-14, 1996.

60.   Hewitt, G.F. and Roberts, D.N., Studies of two-phase flow patterns by simultaneous X-ray and flash photography, UK AEA Report ASRE-M215, 1969.

61.   Hippel, A.R., von, *Dielectric and Waves*, Artech House, Boston, 1995.

62.   Ho, C.M. and Tai, Y.C., Micro-electro-mechanical systems (MEMS) and fluid flows, *Annu. Rev. Fluid Mech.*, vol.30, pp.579-612, 1998.

63.   Holman, J.P., *Heat Transfer*, McGraw-Hill, New York, 1992.

64.   Horiuchi, M. and Aoki, S., Characteristics of silicon wafer-bond strengthening by annealing, *J. Electrochem. Soc.*, vol.139, pp.2589-2594, 1992.

65.   Howe, R.T., Muller, R.S., Gabriel, K.J. and Trimmer, W.S.N., Silicon micromechanics: sensors and actuators on a chip, *IEEE SPECTRUM*, vol.27, pp.29-35, 1990.

66.   Hu, H.Y., Peterson, G.P., Peng, X.F. and Wang, B.X., Interphase fluctuation propagation and superposition model for boiling nucleation, *Int. J. Heat Mass Transfer*, vol.41, pp.3483-3489, 1998.

67.   Huang, J.B., Jiang FK, Tai YC, Ho CM. A micro-electro-mechanical-system-based thermal shear-stress sensor with self-frequency compensation, *Meas. Sci. Technol.*, vol.10, pp.687-96, 1999.

68.   Huff, M.A., Senturia, S.D. and Howe R.T., A thermally isolated microstructure suitable for gas sensing applications, *Solid State Sensor and Actuator Workshop*, *Technical Digest*, pp.47-50, 1988.

69.   Hunter, R.J., *Zeta Potential in Colloid Science, Principles and Applications*, Academic Press, New York, 1981.

70.   Hwang, L., Turlik, I. and Reisman, A., A thermal module design for advanced packaging, *J. Electron. Mater.*, vol.16, pp.347-355, 1987.

71.   Incropera, F.P. and Dewitt, D.P., *Fundamental of Heat and Mass Transfer, 5th ed.*, Wiley, New York, 2002.

72.   International Technology Roadmap for Semiconductors, ITRS 2000 Update, [Online]. Available: http://public.itrs.net/Files/2000Update Final /2kUdFinal.htm

73. Itoh, A. and Polásek, F., Development and application of micro heat pipes, *Proc. 7th Int. Heat Pipe Conf.*, 1990.

74. Jacobi, A.M., Flow and heat transfer in microchannels using a microcontinuum approach, *J. Heat Transfer*, vol.111, pp.1083-1085, 1989.

75. James, E.O. and Carl, M.S., Parallel channel effects for low flow boiling in vertical thin rectangular channels, HTD-vol.273, *Fundamentals of Phase Change: Boiling and Condensation, ASME*, pp.81-91, 1994.

76. Jiang, F., *Ph.D. thesis*, California Institute of Technology, USA, 1997.

77. Jiang, L., *Ph.D. thesis*, Hong Kong University of Science and Technology, Hong Kong, 1999.

78. Jiang, L., Mikkelsen, J., Koo J.M., Huber, D., Yao, S., Zhang, L., Zhou, P., Maveety, J.G., Prasher, R., Santiago, J.G., Kenny, T.W. and Goodson, K.E., Closed-loop electroosmotic microchannel cooling system for VLSI circuits, *IEEE Trans. Components & Packaging Technologies*, in press, 2002.

79. Jiang, F., Tai, Y.C., Ho, C.M. and Li, W.J., A micromachined polysilicon hot-wire anemometer *Solid- State Sensor and Actuator Workshop,* pp.264-267, 1994.

80. Jiang, L., Wang, Y., Wong, M. and Zohar, Y., Fabrication and characterization of a microsystem for microscale heat transfer study, *J. Micromech. Microeng.*, vol.9, pp.422-428, 1999.

81. Jiang, L., Wong, M. and Zohar, Y. Micromachined polycrystalline thin film temperature sensors, *Meas. Sci. Technol.*, vol.10, pp.653-664, 1999.

82. Jiang, L., Wong, M. and Zohar, Y. Phase change in microchannel heat sinks with integrated temperature sensors, *J. Microelectromech. Syst.*, vol.8, pp.358-365, 1999.

83. Jiang, L., Wong, M. and Zohar, Y., Unsteady characteristics of a thermal micro system, *Sensors Actuators A*, vol.82, pp.108-113, 2000.

84. Jiang, L., Wong, M. and Zohar, Y., Transient temperature performance of an integrated thermal microsystem, *J. Micromech. Microeng.*, vol.10, pp.466-476, 2000.

85. Jiang, L., Wong, M. and Zohar, Y., Forced convection boiling in a microchannel heat sink, *J. Microelectromech. Syst.*, vol.10, pp.80-87, 2001.

86. Kandlikar, S.G., Fundamental issues related to flow boiling in minichannels and microchannels, *Exp. Thermal Fluid Sci.*, vol.26, pp.389-407, 2002.

87. Kandlikar, S.G., Mizo, V.R. and Cartwright, M.D., Investigation of bubble departure mechanism in subcooled flow boiling of water using

high-speed photography, in *Convective flow boiling*, J.C. Chen, Ed., Taylor & Francis, Washington, DC, pp.161-166, 1996.

88. Kandlikar, S.G., Mizo, V., Cartwright, M. and Lkenze, E., Bubble nucleation and growth characteristics in subcooled flow boiling of water, ASME HTD-vol.342, *National Heat Transfer Conference*, vol.4, pp.11-18, 1997.

89. Katto, Y., An analysis of the effect of inlet subcooling on critical heat flux of forced convection boiling in vertical uniformly heated tubes, *Int. J. Heat Mass Transfer,* vol.22, pp.1567-1575, 1979.

90. Kavehpour, H.P., Faghri, M. and Asako, Y., Effects of compressibility and rarefaction on gaseous flows in microchannels, *Numer. Heat Transfer A*, vol.32, pp.677-695, 1997.

91. Kennard, E.H., *Kinetic Theory of Gases*, McGraw-Hill, New York, 1938.

92. Kennedy, J.E., Roach, G.M., Jr., Dowling, M.F., Abdel-Khalik, S.I., Ghiaasiaan, S.M., Jeter, S.M., and Quereshi, Z.H., The onset of flow instability in uniformly heated horizontal microchannels, *J. Heat Transfer*, vol.122, pp.118-125, 2000.

93. Keyes, R.W., Heat transfer in forced convection through fins, *IEEE Trans. Electron Devices*, vol.ED-31, pp.1218-1221, 1984.

94. Khrustalev, D., and Faghri, A., Thermal analysis of a micro heat pipe, *J. Heat Transfer*, vol.116, pp.189–198, 1994.

95. Kishimoto, T. and Ohsaki, T., VLSI packaging technique using liquid-cooled channels, *36th Electronic Components Conference Proceedings, IEEE,* pp.595-601, 1986.

96. Klausner, J.F. and Mei, R., Suppression of nucleation sites in flow boiling, in *Convective flow boiling*, J.C. Chen, Ed., Taylor & Francis, Washington, DC, pp.155-160, 1996.

97. Kleiner, M.B., Kuehn, S.A. and Haberger, K., High performance forced air cooling scheme employing microchannel heat exchangers, *IEEE Trans. Components, Packaging, Manuf. Technol. A*, vol.18, pp.795–804, 1995.

98. Knight, R.W., Goodling, J.S. and Hall, D.J., Optimal thermal design of forced convection heat sinks-analytical, *J. Electron. Packaging,* vol.113, pp.313-321, 1991.

99. Koh, J.C.Y. and Colony R., Heat transfer of microstructures for integrated circuits, *Int. Comm. Heat Mass Transfer*, vol. 13, pp.89-98, 1986.

100. Kohl, F., Jachimowicz, A., Steurer, J., Glatz, R., Kuttner, J., Biacovsky, D. and Olcaytug, F., A micromachined flow sensor for liquid and gaseous fluids, *Sensors Actuators A*, vol.41, pp.293-299, 1994.

186

101. Kopp, M.U., deMello, A.J. and Manz, A., Chemical amplification: continuous-flow PCR on a chip, *Science*, vol.280, pp.1046-1048, 1998.
102. Korsh, G.J. and Muller, R.S., Conduction properties of lightly doped polycrystalline silicon, *Solid State Electronics*, vol.21, pp.1045-1051, 1978.
103. Kun, T.K. and Kim, C.J., Valveless pumping using traversing vapor bubbles in microchannels, *J. Appl. Phys.*, v.83, pp.5658-5664, 1998.
104. Kuttner, H., Urban, G., Jachimowicz, A., Kohl, F., Olcaytug, F. and Goiser, P., Microminiaturized thermistor arrays for temperature gradient, flow and perfusion measurements, *Sensors Actuators A*, vol.27, pp.641-645, 1991.
105. Lazarek, G.M. and Black, S.H., Evaporative heat transfer, pressure drop and critical heat flux in a small vertical tube with R-113, *Int. J. Heat Mass Transfer*, vol.25, pp.945-960, 1982.
106. Lee, D. Y. and Vafai, K., Comparative analysis of jet impingement and microchannel cooling for high heat flux applications, *Int. J. Heat Mass Transfer*, vol. 42, pp. 1555–1568, 1999.
107. Lee, M., Wong, Y.Y., Wong, M. and Zohar, Y., Two-phase flow boiling in a microchannel heat sink. *International Mechanical Engineering Congress & Exposition*, Paper IMECE2001/MEMS-23881, 2001.
108. Lee, M., Wong, Y.Y., Wong, M. and Zohar, Y., Size and shape effects on two-phase flow instabilities in microchannels, *Proc.15th Int. Conf. Micro Electro Mechanical Systems, MEMS'02*, pp.28-31, 2002.
109. Lee, M., Wong, M. and Zohar, Y., Design, fabrication and characterization of an integrated micro heat pipe, *Proc.15th Int. Conf. Micro Electro Mechanical Systems, MEMS'02*, pp.85-88, 2002.
110. Lee, K.T. and Yan, W.M., Transient conjugated forced convection heat transfer with fully developed laminar flow in pipes, *Numer. Heat Transfer A*, vol.23, pp.341-359, 1993.
111. Lide, D.R., Ed., *Handbook of Chemistry and Physics*, CRC Press, Boca Raton, FL, 1995.
112. Little, W.A., Advances in Joule-Thomson cooling, *Advances in Cryogenic Engineering*, vol.35, pp.1305-1314, 1990.
113. Liu, J.Q., Tai Y.C., Pong, K.C. and Ho, C.M., Micromachined channel/pressure sensor systems for micro flow studies, *Proc. 7th Int. Conf. Solid-State Sensors Actuators, Transducers'93*, pp.995-997, 1993.
114. Longtin, J.P., Badran, B. and Gerner, F.M., A one-dimensional model of a micro heat pipe during steady-state operation, *J. Heat Transfer*, vol.116, pp.709-715, 1994.

115. Lord, R., Comparative measurements of tangential momentum and thermal accommodations on polished and on roughened steel spheres, in *Rarefied Gas Dynamics*, K. Karamcheti, Ed., Academic, New York, 1974.

116. Lu, L., Gerzberg, N.C.C. and Meindl, J.D., A quantitative model of the effect of grain size on the resistivity of polycrystalline silicon resistors, *IEEE Electron Device Letters*, vol.1, pp.38-41, 1980.

117. Lukaszewicz, G. *Micropolar Fluids: Theory and Applications*, Birkhauser, Boston, 1999.

118. Ma, H.B. and Peterson, G.P., Laminar friction factor in microscale ducts of irregular cross section, *Microscale thermophys. Eng.*, vol.1, pp.253-265, 1997.

119. Ma, H.B., Peterson, G.P. and Lu, X.J., The influence of vapor-liquid interactions on the liquid pressure drop in triangular microgrooves, *Int. J. Heat Mass Transfer*, vol.37, pp.2211-2219, 1994.

120. Mahalingam, M., Thermal management in semiconductor device packaging, *Proc. IEEE*, vol.73, pp.1396-1404, 1985.

121. Makino, E., Mitsuya, T. and Shibata, T., Fabrication of TiNi shape memory micropump, *Sensors Actuators A*, vol.88, pp.256-262, 2001.

122. Mala, G.M., Li, D. and Dale, J.D., Heat transfer and fluid flow in microchannels, *Int. J. Heat Mass Transfer*, vol.40, pp.3079-3088, 1997.

123. Mallik, A.K. and Peterson, G.P., Steady-state investigation of vapor deposited micro heat pipe arrays, *J. Electron. Packaging*, vol. 117, pp.75–81, 1995.

124. Mallik, A.K., Peterson, G.P. and Weichold, M.H., Construction processes for vapor deposited micro heat pipes, *10th Symp. Electronic Materials Processing and Characteristics*, 1991.

125. Mallik, A.K., Peterson, G.P. and Weichold, M.H., On the use of micro heat pipes as an integral part of semiconductor devices, *J. Electron. Packaging*, vol.114, pp.436–442, 1992.

126. Marongiu, M.J., Compressibility effects in the design of gas-cooled microchannel heat sinks, *Proc. Intersociety Conf. Thermal Phenomena Electronic Systems*, pp.124-131, 1996.

127. Mastrangelo, C., *Ph.D. thesis,* University of California at Berkeley, Berkeley, CA, 1990.

128. Mavriplis, C., Ahn, J.C. and Goulard, R., Heat transfer and flowfields in short microchannels using direct simulation Monte Carlo, *J. Thermophys. Heat Transfer*, vol.11, pp.489-496, 1997.

129. Mayer, F., Salis, G., Funk, J., Paul, O. and Baltes, H., Scaling of thermal CMOS gas flow microsensors: experiment and simulation,

Proc. 9th Int. Workshop Micro Electro Mechanical Systems,
MEMS'96, pp.116-121, 1996.

130. Mills, A.F, *Heat Transfer,* Prentice Hall, Upper Saddle River, N.J., 1999.

131. Moffat, R.J., Arvizu, D.E., and Ortega, Cooling electronic components: forced convection experiments with an air-cooled array, ASME HTD-vol.48, pp.5-15, 1985.

132. Morega, A.M. and Bejan, A., Optimal spacing of parallel boards with discrete heat sources cooled by laminar forced convection, *Numer. Heat Transfer A,* vol.25, pp.373-392, 1994.

133. Morega, A.M., Vargas J.V.C. and Bejan A., Optimization of pulsating heaters in forced convection, *Int. J. Heat Mass Transfer,* vol.38, pp.2925-2934, 1995.

134. Mudawar, I. and Bowers, M.B., Ultra-high critical heat flux (CHF) for subcooled water flow boiling-I: CHF data and parametric effects for small diameter tubes, *Int. J. Heat Mass Transfer,* vol.42, pp.1405-1428, 1999.

135. Munch, U., Jaeggi, D., Schneeberger, K., Schaufelbuhl, A., Paul, O., Baltes, H. and Jasper, J., Industrial fabrication technology for CMOS infrared sensor arrays, *Proc. 9th Int. Conf. Solid-State Sensors Actuators, Transducers'97,* pp.205-208, 1997.

136. Nayak, D., Hwang, L., Turlik, I. and Reisman, A., A high-performance thermal module for computer packaging, *J. Electron. Mater.,* vol.16, pp.357-364, 1987.

137. Northrup, M.A., Ching, M.T., White, R.M. and Watson, R.T., DNA amplification with a microfabricated reaction chamber, *Proc. 7th Int. Conf. Solid-State Sensors Actuators, Transducers'93,* pp.924-926, 1993.

138. Northrup, M.A., Gonzalez, C., Hadley, D., Hills, R.F., Landre, P., Lehew, S., Saiki, R., Sninsky, J. J., Watson, R. and Watson, R., Jr., A MEMS-based miniature DNA analysis system, *Proc. 10th Int. Conf. Solid-State Sensors Actuators, Transducers'95,* pp.764-765, 1995.

139. Ortiz, L., Gonzalez, J.E., Experiments on steady-state high heat fluxes using spray cooling, *Exp. Heat Transfer,* vol.12, pp.215-233, 1999.

140. Palm, B., Heat transfer in microchannels, *Microscale thermophys. Eng.,* vol.5, pp.155-175, 2001.

141. Papautsky, I., Brazzle, J., Ameel, T. and Frazier, A.B., Laminar fluid behavior in microchannels using micropolar fluid theory, *Sensors Actuators A,* vol.73, pp.101-108, 1999.

142. Park, J.S., Chu, L.L., Oliver, A.D. and Gianchandani, Y.B., Bent-beam electrothermal actuators-Part II: Linear and rotary microengines, *J. Microelectromech. Syst.,* vol.10, pp.255-262, 2001.

143. Peeters, E., Verdonckt-Vandebroek, S., Thermal ink jet technology, *IEEE Circuits & Devices Magazine*, v.13, pp.19-23, 1997.

144. Peles, Y.P. and Haber, S., A steady state, one dimensional model for boiling two-phase flow in triangular micro-channel, *Int. J. Multiphase Flow*, vol.26, pp.1095-1115, 2000.

145. Peles, Y.P., Yarin, L.P. and Hetsroni, G., Thermodynamic characteristics of two-phase flow in a heated capillary, *Int. J. Multiphase Flow*, vol.26, pp.1063-1093, 2000.

146. Peles, Y.P., Yarin L.P. and Hetsroni G., Steady and unsteady flow in a heated capillary, *Int. J. Multiphase Flow*, vol.27, pp.577-598, 2001.

147. Pelisser, S., Pigeon, F., Biasse, B., Zussy, M., Pandraud, G. and Mure-Revaud, A., New technique to produce buried channel waveguides in glass, *Optical Engineering*, vol.37, pp.1111-1114, 1998.

148. Peng, X.F., and Peterson, G.P., The effect of thermofluid and geometrical parameters on convection of liquids through rectangular microchannels, *Int. J. Heat Mass Transfer*, vol.38, pp.755-758, 1995.

149. Peng, X.F. and Peterson, G.P., Convective heat transfer and flow friction for water flow in microchannel structures, *Int. J. Heat Mass Transfer*, vol.39, pp.2599-2608, 1996.

150. Peng, X.F. and Peterson, G.P., Forced convection heat transfer of single-phase binary mixtures through microchannels, *Exp. Thermal Fluid Sci.*, vol.12, pp.98-104, 1996.

151. Peng, X.F., Peterson, G.P. and Wang, B.X., Frictional flow characteristics of water flowing through rectangular microchannels, *Exp. Thermal Fluid Sci.*, vol.7, pp.249-264, 1994.

152. Peng, X.F., Peterson, G.P. and Wang, B.X., Heat transfer characteristics of water flowing through microchannels, *Exp. Heat Transfer*, vol.7, pp.265-283, 1994.

153. Peng, X.F., Peterson, G.P. and Wang, B.X., Flow boiling of binary mixtures in microchanneled plates, *Int. J. Heat Mass Transfer*, vol.39, pp.1257-1264, 1996.

154. Peng, X.F. and Wang, B.X., Forced convection and flow boiling heat transfer for liquid flowing through microchannels, *Int. J. Heat Mass Transfer*, vol.36, pp.3421-3427, 1993.

155. Peng, X.F., Wang, B.X., Peterson, G.P. and Ma, H.B, Experimental investigation of heat transfer in flat plates with rectangular microchannels, *Int. J. Heat Mass Transfer*, vol.38, pp.127-137, 1995.

156. Petersen, K.E., Silicon as a mechanical material, *Proc. IEEE*, vol.70, pp.420-457, 1982.

157. Petersen, K., Biomedical applications of MEMS, *Int. Electron Devices Meeting, Technical Digest*, pp.239-242, 1996.

190

158. Peterson, G.P., An overview of micro heat pipe research, *Appl. Mechanics Rev.*, vol.45, pp. 175–189, 1992.

159. Peterson, G.P., *An Introduction to Heat Pipes: Modeling, Testing and Applications,* Wiley, New York, 1994.

160. Peterson, G.P., Micro heat pipes and micro heat spreaders, in *The CRC Handbook of MEMS*, M. Gad-el-Hak, Ed., CRC Press, Boca Raton, Florida, pp. 31:1-26, 2001.

161. Peterson, G.P., Duncan, A.B., and Weichold, M.H., Experimental investigation of micro heat pipes fabricated in silicon wafers, *J. Heat Transfer*, vol.115, pp.751–756, 1993.

162. Peterson, G.P. and Ma, H.B., Theoretical analysis of the maximum heat transport in triangular grooves: A study of idealized micro heat pipes, *J. Heat Transfer*, vol.118, pp.731–739, 1996.

163. Peterson, G.P. and Mallik, A.K., Transient response characteristics of vapor deposited micro heat pipe arrays, *J. Electron. Packaging*, vol.117, pp.82–87, 1995.

164. Peterson, G.P., Swanson, L.W. and Gerner, F.M., Micro heat pipes, in *Microscale Energy Transport*, C.L. Tien, A. Majumdar and F.M. Gerner, Eds., Taylor & Francis, Washington, DC, pp.295-337, 1998.

165. Pettigrew, K., Kirshberg, J., Yerkes, K., Trebotich, D., and Liepmann, D., Performance of a MEMS based micro capillary pumped loop for chip-level temperature control, *Proc. 14th Int. Conf. Micro Electro Mechanical Systems, MEMS'01*, pp.427-430, 2001.

166. Pfahler, J., Harley, J.C., Bau, H.H. and Jemel, J.N., Liquid transport in micron and submicron channels, *Sensors Actuators A*, vol.22, pp.431-434, 1990.

167. Pong, K.C., Ho, C.M., Liu, J. and Tai, Y.C., Non-linear pressure distribution in uniform microchannels, ASME FED-vol.197, pp.51-56, 1994.

168. Que, L., Park, J.S. and Gianchandani, Y.B., Bent-beam electrothermal actuators-Part I: Single beam and cascaded devices, *J. Microelectromech. Syst.*, vol.10, pp.247-254, 2001.

169. Quenzer, H.J. and Benecke, W., Low-temperature silicon wafer bonding, *Sensors Actuators A*, vol.32, pp.340-344, 1992.

170. Rachkovsfij, D.A., Kussul, E.M. and Talayev, S.A., Heat exchanger in short microtubes and micro heat exchangers with low hydraulic losses, *Microsyst. Technol.*, vol.4, pp.151-158, 1998.

171. Rahman, M.M. and Gui. F., Experimental measurements of fluid flow and heat transfer in microchannel cooling passages in a chip substrate, *Adv. Electron. Packaging*, ASME EEP-vol.2, pp.685–692, 1993.

172. Ramaswamy, C., Joshi, Y., Nakayama, W. and Johnson, W.B., Compact thermosyphons employing microfabricated components, *Microscale thermophys. Eng.*, vol.3, pp.273-282, 1999.

173. Ravigururajan, T.S., Impact of channel geometry on two-phase flow heat transfer characteristics of refrigerants in microchannel heat exchangers, *J. Heat Transfer*, vol.120, pp.485-491, 1998.

174. Remtema, T. and Lin, L., Active frequency tuning for microresonators by localized thermal stressing effects, *Solid State Sensor and Actuator Workshop, Technical Digest*, pp.363-366, 2000.

175. Roach, G.M. Jr., Abdel-Khalik, S.I., Ghiaasiaan, S.M., Dowling, M.F. and Jeter, S.M., Low-flow critical heat flux in heated microchannels, *Nuclear Sci. Eng.*, vol.131, pp.411–425, 1999.

176. Rujano, J.R. and Rahman, M.M., Transient response of microchannel heat sinks in a silicon wafer, *J. Electron. Packaging*, vol.119, pp.239-246, 1997.

177. Sadik, K. and Yaman, Y., *Convective Heat Transfer, 2nd ed.*, CRC Press, Boca Raton, Florida, 1995.

178. Samalam, V.K., Convective heat transfer in microchannels, *J. Electron. Mater.*, vol.18, pp.611-617, 1989.

179. Schaaf, S.A. and Chambre, P.L., *Flow of Rarefied Gases*, Princeton University Press, Princeton, New Jersey, 1961.

180. Schutte, D.J., Rahman, M.M. and Faghri, A., Transient conjugate heat transfer in a thick-wall pipe with developing laminar flow, *Numer. Heat Transfer A*, vol.21, pp.163-186, 1992.

181. Serizawa, A. and Feng, Z.P., Two-phase flow in microchannels, *Proc. 4th Int. Conf. Multiphase Flow*, 2001.

182. Shah, R.K. and Bhatti, M.S., Laminar convective heat transfer in ducts, in *Handbook of Single Phase Convective Heat Transfer*, Eds. S. Kakac, R.K. Shah and W. Aung, Chapter 3, Wiley, New York, 1987.

183. Shoffner, M.A., Cheng, J., Hvichia G.E., Kricka L.J. and Wilding P., Chip PCR. I. Surface passivation of microfabricated silicon-glass chips for PCR, *Nucleic Acids Res.*, vol.24, pp.375-379, 1996.

184. Shoji, S., Kikuchi, H. and Torigoe, H., Low-temperature anodic bonding using lithium aluminosilicate-β-quartz glass ceramic, *Sensors Actuators A*, vol.64, pp.95-100, 1998.

185. Sobhan, C.B. and Garimella, S.V., A comparative analysis of studies on heat transfer and fluid flow in microchannels, *Microscale thermophys. Eng.*, vol.5, pp.293-311, 2001.

186. Sone, Y., Aoki, K., Takata, K., Sugimoto, H. and Bobylev, A.V., Inappropriateness of the heat-conduction equation for description of a temperature field of a stationary gas in the continuum limit:

Examination by asymptotic analysis and numerical computation of the Boltzmann equation, *Phys. Fluids*, vol.8, pp.628-638, 1996.

187. Sone, Y., Waniguchi, Y. and Aoki, K., One-way flow of a rarefied gas induced in a channel with a periodic temperature distribution, *Phys. Fluids*, vol.8, pp.2227-2235, 1996.

188. Sparrow, E.M., Baliga, B.R. and Patankar, S.V., Forced convection heat transfer from a shrouded fin array with and without tip clearance, *J. Heat Transfer*, vol.100, pp.572-579, 1978.

189. Stanley, R. S., Ameel, T. A. and Warrington, R. O., Convective flow boiling in microgeometries: a review and application, in *Convective flow boiling*, J.C. Chen, Ed., Taylor & Francis, Washington, DC, pp. 305-310, 1996.

190. Stephan, K., *Heat Transfer in Condensation and Boiling*, Springer-Verlag, Berlin, 1992.

191. Stokes, V.K., *Theories of Fluids with Microstructure*, Springer-Verlag, Berlin, 1984.

192. Tai, Y.C., Mastrangelo, C.H. and Muller, R.S., Thermal conductivity of LPCVD polycrystalline silicon, *J. Appl. Physics,* vol.63, pp. 1442-1447, 1988.

193. Tai, Y.C. and Muller, R.S., Lightly doped polysilicon bridge as a flow meter, *Sensors Actuators,* vol.15, pp.63-75, 1988.

194. Tai, Y.C., Muller R.S. and Howe, R.T., Polysilicon-bridges for anemometer applications, *Proc. 3rd Int. Conf. Solid-State Sensors Actuators, Transducers'85*, pp.354-357, 1985.

195. Taitel, Y. and Dukler, A.E., A model for predicting flow regimes transitions in horizontal and near horizontal gas-liquid flow, *AIChE J.*, vol.22, pp.47-55, 1976.

196. Tang, L. and Joshi, Y.K., Integrated thermal analysis of indirect air-cooled electronic chassis, *IEEE Trans. Components, Packaging, Manuf. Technol. A*, vol.20, pp.103–110, 1997.

197. Taylor, T.B., Winn-Deen, E.S., Picozza, E., Woudenberg, T.M. and Albin M., Optimization of the performance of the polymerase chain reaction in silicon-based microstructures, *Nucleic Acids Res.*, vol.25, pp.3164-3168, 1997.

198. Tien, C.L. and Kuo, S.M., Analysis of forced convection in microstructures for electronic system cooling, in *Cooling Technology for Electronic Equipment*, W. Aung, Ed., Hemisphere Pub. Corp., New York, pp.285-294, 1988.

199. Time, R.W., A field-focusing capacitance sensor for multiphase flow analysis, *Sensors Actuators A*, vol.21, pp.115-122, 1990.

200. Trimmer, W.S.N., Microrobots and micromechanical systems, *Sensors Actuators*, vol.19, pp.267-287, 1989.

201. Triplett, K.A., Ghiaasiaan, S.M., Abdel-Khalik, S.I. and Sadowski, D.L., Gas-liquid two phase flow in microchannels. Part I: two phase flow patterns, *Int. J. Multiphase Flow*, vol.25, pp.377–394, 1999.
202. Triplett, K.A., Ghiaasiaan, S.M., Abdel-Khalik, S. I., LeMouel, A. and McCord, B.N, Gas-liquid two phase flow in microchannels. Part II: void fraction and pressure drop, *Int. J. Multiphase Flow*, vol.25, pp.395–410, 1999.
203. Tsai, J.H. and Lin, L., A thermal bubble actuated micro nozzle-diffuser pump, *Proc.14th Int. Conf. Micro Electro Mechanical Systems, MEMS'01*, pp.409-412, 2001.
204. Tseng, F.G., Kim, C.J., Ho, C.M., A novel micro injector with virtual chamber neck, *Proc. 11th Int. Workshop Micro Electro Mechanical Systems, MEMS'98*, pp.57-62, 1998.
205. Tso, C.P. and Mahulikar, S.P., Use of the Brinkman number for single phase forced convective heat transfer in microchannels, *Int. J. Heat Mass Transfer*, vol.41, pp.1759–1769, 1998.
206. Tso, C.P. and Mahulikar, S.P., Role of the Brinkman number in analyzing flow transitions in microchannels, *Int. J. Heat Mass Transfer*, vol.42, pp.1813–1833, 1999.
207. Tso, C.P. and Mahulikar, S.P., Experimental verification of the role of Brinkman number in microchannels using local parameters, *Int. J. Heat Mass Transfer*, vol.43, pp.1837–1849, 2000.
208. Tuckerman, D.B. and Pease, R.F.W., High-performance heat sinking for VLSI, *IEEE Electron Device Lett*, vol.2, pp.126-129, 1981.
209. Tuckerman, D.B. and Pease, R.F.W., Optimised convective cooling using micromachined structures, *Electrochem. Soc. Ext. Abstract*, vol.82, pp.197-198, 1982.
210. Unal, C. and Pasamehmetoglu, K.O., Spatial and temporal variation of the surface temperature and heat flux for saturated pool nucleate boiling at lower heat fluxes, in *Fundamentals of Phase Change: Boiling and Condensation*, ASME HTD-Vol.273, pp.49-56, 1994.
211. Urban, G., Jachimovwicz, A., Kohl, F., Kuttner, H., Olcaytug, F., Goiser, P., Lindner, K., Pockberger, H., Prohaska, O. and Schonauer, M., High resolution multi-temperature sensors for biomedical applications, *Medical Progress through Technology*, vol.16, pp.173-181, 1990.
212. Wang, C.H. and Dhir, V.K., On the gas entrapment and nucleation site density during pool boiling of saturated water, *J. Heat Transfer*, vol.115, pp.670-679, 1993.
213. Wang, B.X. and Peng, X.F., Experimental investigation on liquid forced-convection heat transfer through microchannels, *Int. J. Heat Mass Transfer*, vol.37, suppl.1, pp.73-82, 1994.

194

214. Wang, T.C., Snyder, T.J. and Chung, J.N., Forced-convection subcooled nucleate boiling and its application in microgravity, in *Convective flow boiling*, J.C. Chen, Ed., Taylor & Francis, Washington, DC, pp.111-116, 1996.

215. Wang, R.X., Zohar, Y. and Wong, M., Residual stress loaded titanium-nickel shape-memory alloy thin-film micro-actuators, *J. Micromech. Microeng.*, vol.12, pp.323-327, 2002.

216. Waters, L.C., Jacobson, S.C., Kroutchinina, N., Khandurina, J., Foote, R.S. and Ramsey, J.M., Microchip device for cell lysis, multiplex PCR amplification, and electrophoretic sizing, *Anal. Chem.*, vol.70, pp.158-162, 1998.

217. Weisberg, A., Bau, H.H. and Zemel, J.N., Analysis of microchannels for integrated cooling, *Int. J. Heat Mass Transfer*, vol.35, pp.2465-2474, 1992.

218. White, F.M., *Heat and Mass Transfer*, Addison-Wesley, Reading, MA, 1991.

219. Williams M.M.R., Boundary-value problems in the kinetic theory of gases. Part I. Slip flow, *J. Fluid Mech.* vol.36, pp.145-159, 1969.

220. Williams M.M.R., Boundary-value problems in the kinetic theory of gases. Part II. Thermal creep, *J. Fluid Mech.* vol.45, pp.759-768, 1971.

221. Wise, K.D. and Najafi, K., Microfabrication techniques for integrated sensors and microsystems, *Science*, vol.254, pp.1335-1342, 1991.

222. Wood, M.H., Muzik, A., Huston, H.H. and Hazara, R., Burn-in, in *Microelectronics Manufacturing Diagonstics Handbook*, A.H. Landzberg, Ed., Van Nostrand Reinhold, NY, 1993.

223. Woolley, A.T., Hadley, D., Landre, P., deMello, A.J., Mathies, R.A. and Northrupt, M.A., Functional integration of PCR amplification and capillary electrophoresis in a microfabricated DNA analysis device, *Anal. Chem.*, vol.68, pp.4081-4086, 1996.

224. Wu, S., *Ph.D. thesis*, California Institute of Technology, USA, 2000.

225. Wu, S., Lin, Q., Yuen, Y. and Tai, Y.C., MEMS flow sensors for nano-fluidic applications, *Sensors Actuators A*, vol.89, pp.152-158, 2001.

226. Wu, P. and Little, W.A., Measurement of the heat transfer characteristics of gas flow in fine channel heat exchangers used for microminiature refrigerators, *Cryogenics*, vol.24, pp.415-420, 1984.

227. Wu, S., Mai, J., Tai, Y.C. and Ho, C.M., MEMS heat exchanger by using MEMS impinging jets, *Proc. 12th Int. Micro Electro Mechanical Systems Conf., MEMS'99*, pp.171-176, 1999.

228. Wu, S., Mai, J., Zohar, Y., Tai, Y.C. and Ho, C.M., A suspended microchannel with integrated temperature sensors for high-pressure

flow studies, *Proc. 11th Int. Workshop Micro Electro Mechanical Systems, MEMS'98*, pp.87-92, 1998.

229. Wu, D. and Peterson, G.P., Investigation of the transient characteristics of a micro heat pipe, *J. Thermophys. Heat Transfer*, vol.5, pp.129–134, 1991.

230. Wu, D., Peterson, G.P. and Chang, W.S., Transient experimental investigation of micro heat pipes, *J. Thermophys. Heat Transfer*, vol.5, pp.539–545, 1991.

231. Yang, C., Li, D. and Masliyah, J.H., Modeling forced liquid convection in rectangular microchannels with electrokinetic effects, *Int. J. Heat Mass Transfer*, vol.41, pp.4229–4249, 1998.

232. Yao, D.J., Kim, C.J. and Chen, G., Design of thin film thermoelectric microcooler, ASME HTD-vol.366-2, pp.345-351, 2000.

233. Yuan, H. and Prosperetti, A., The pumping effect of growing and collapsing bubbles in a tube, *J. Micromech. Microeng.*, vol.9, pp.402-413, 1999.

234. Zahn, J.D., Deshmukh, A.A., Pisano, A.P. and Liepmann, D., Continuous on-chip micropumping through a microneedle, *Proc.14th Int. Conf. Micro Electro Mechanical Systems, MEMS'01*, pp.503-506, 2001.

235. Zhang, L., Koo, J.M., Jiang, L., Asheghi, M., Goodson, K.E., Santiago, J.G. and Kenny, T.W., Measurements and modeling of two-phase flow in microchannels with nearly constant heat flux boundary conditions, *J. Microelectromech. Syst.*, vol.11, pp.12-19, 2002.

236. Zhuang, Y., Ma, C.F. and Qin, M., Experimental study on local heat transfer with liquid impingement flow in two-dimensional micro-channels, *Int. J. Heat Mass Transfer*, vol.40, pp.4055–4059, 1997.

237. Zhukov, V.M. and Yarmak, I.L., Transient heat transfer in two-phase cryogenic liquid forced flows under step heat flux in narrow channels, *Cryogenics*, vol.30, pp.282-286, 1990.

238. Zohar, Y., Microchannel heat sinks, in *The CRC Handbook of MEMS*, M. Gad-el-Hak, Ed., CRC Press, Boca Raton, Florida, pp. 32:1-30, 2001.

239. Zohar, Y., Chu, W.K.H., Hsu, C.T. and Wong, M., Slip flow effects on heat transfer in microchannels, *Bull. Am. Phys. Soc.*, vol.39, p.1908, 1994.

240. Zohar, Y., Lee, W.Y., Lee, S.Y.K., Jiang, L. and Tong, P., Subsonic gas flow in a straight and uniform microchannel, *J. Fluid Mech.*, in print, 2002.

241. Zou, J., Ye, X.Y., Zhou, Z.Y. and Yang, Y., A novel thermally-actuated silicon micropump, *Proc. Int. Symp. Micromech. & Human Sci.*, pp.231-234, 1997.

242. Zumbrunnen, D.A., Transient convection heat transfer in planar stagnation flow with time-varying surface heat flux and temperature, *J. Heat Transfer*, vol.114, pp.85-93, 1992.

Index